Courses, Components, and Exercises in Technical Communication

Courses, Components, and Exercises in Technical Communication

Edited by

Dwight W. Stevenson
The University of Michigan

with

Renee R. Betz
University of Illinois at Chicago Circle

David L. Carson
Rensselaer Polytechnic Institute

Donald H. Cunningham
Morehead State University

Thomas M. Sawyer
The University of Michigan

National Council of Teachers of English
1111 Kenyon Road, Urbana, Illinois 61801

Book Design: Tom Kovacs

NCTE Stock Number 08776

Library of Congress Cataloging in Publication Data

Main entry under title:

Courses, components, and exercises in technical communication.

    Bibliography: p.
    1. Technical writing—Addresses, essays, lectures.
I. Stevenson, Dwight W., 1933–   . II. National
Council of Teachers of English.
T11.C68   808'.0666021   81-4002
ISBN 0-8141-0877-6   AACR2

Contents

Preface

Technical writing is a rapidly growing and developing discipline in which teachers new to the field as well as those already experienced in it are seeking new professional resources. This collection of articles on teaching methods and assignments provides one such new source of help.

The collection is new in three respects. First, the articles presented here were written specifically for this book. They were not, as is so often the case with anthologies, written in response to varied calls for papers or for oral presentation at professional conferences. Second, the articles in this anthology were written in such a way as to make them particularly helpful to technical writing teachers in the classroom. They are detailed in their explanations of methodology and often include illustrative material and discussions of evaluation techniques. Third, and most important, the contributors to this anthology have made special effort to advise teachers on how to make classroom activities directly relevant to writing done in the world beyond the classroom—the world of business, industry, and government. Students taught by methods suggested in this collection should be unusually well prepared to assume the writing tasks they will encounter in their careers beyond graduation. They should be especially sensitive to the implications of audiences and purposes similar to those with which they will deal as professionals. Further, they should be aware that communicating well as a professional in industry, business, or government demands a variety of skills, including research skills, graphic communication skills, oral communication skills, and group writing skills. They should see writing as a complex skill they must master if they are to function effectively in the professional world.

The anthology is divided into three major sections. The first, Courses, presents plans for alternative approaches to the teaching of technical writing. The second, Components, offers a variety of activities for major segments within a course. And the third, Exercises, suggests individual activities that can be completed during a few class periods within a course or component. Thus teachers searching for general approaches and strategies as well as those looking for specific activities to incorporate within the structure of their present course will find this collection valuable. Obviously not intended to be exhaustive, the collection should nonetheless help technical writing teachers meet the increasing demand for their courses, the higher expectations of their departments and students, and the more and more complex communication needs of our society.

Acknowledgments

This anthology is the result of a project initiated by the National Council of Teachers of English Committee on Technical and Scientific Communication, and thus to the members of that committee thanks are due. It was they who sponsored the call for manuscripts and who made suggestions on appropriate content. Yet the committee members are more than administratively responsible for this anthology; they are a part of this anthology in the sense that they offer their services as human resources to complement the written resources provided by the articles in the collection. Thus, the committee page in this anthology both identifies those whose efforts should be recognized and offers their further services to users of the collection.

One final personal word of thanks. Two individuals must be singled out as being particularly important in their contributions to this project. These are Renee Betz and Donald H. Cunningham. Renee worked many hours on the manuscript. Don kept pushing until the collection was finished. Without their encouragement and work the anthology would not have been done. I deeply appreciate their efforts.

Dwight W. Stevenson
The University of Michigan

Part One: Courses

1 A Professional Scenario for the Technical Writing Classroom

Lawrence J. Johnson
University of Texas at El Paso

Through a series of seven assignments completed in support of research in their career fields, students learn skills that will help them communicate with the varied and demanding audiences they will encounter in professional life. Classroom work includes the identification of structures appropriate to given purposes and audiences and the analysis of student papers.

At my university, a single technical writing course (three semester credit hours) is required of students majoring in the sciences, in metallurgical engineering, and in criminal justice. These students may take this course immediately upon completion of freshman composition, or they may wait until as late as the last semester of their senior year. Consequently, each class enrolls a wide range of students: while some students are concurrently enrolled in a senior seminar, others have yet to enroll in their first nonsurvey science course; while some plan to go on to graduate school, others will seek employment with a baccalaureate degree. Five years ago, we redesigned this course; in doing so, we attempted to develop an approach that would serve students, whatever their majors, their experiences, and their ambitions. The design centers on seven assignments that may be completed in support of a research project directed by a professor from the student's major field or within the context of a classroom scenario that specifies an audience and a problem to be addressed by the student.

Rationale

We know that every professional field prescribes distinctive forms and formats for its writers, but we also know that each profession

3

rapidly instructs new graduates in the idiosyncrasies of its particular technical writing repertory. We did not, therefore, see this technical writing course as a course in writing for any particular profession, though we intended to help students learn to recognize the legitimate applications of whatever repertory they might encounter. Instead, we designed our technical writing course to teach students how to respond effectively to a variety of written demands from intelligent audiences who, for some specific reason, need specific information.

We have found that professionals tend to have difficulty when they are called upon to communicate specialized knowledge to intellectual peers in disciplines other than their own. Contrary to popular belief, most professionals spend as much time and effort writing to lay audiences as they do in formal communication with co-workers. This is especially true outside of academia, where modern organizational structures demand extensive reporting both up and down managerial ladders. It is the ability to meet such demands that we try to develop in our technical writing students.

This technical writing course is further shaped by our belief that a professional's primary asset is the information he or she possesses but that such information is of value only when it is successfully communicated to someone who needs it. Any attempt to communicate is potentially frustrating: the nature of the information, the predisposition of the audience, the rhetorical choices made by the writer—each may militate against successful communication. Too many students have managed to evade rather than to work through such frustration by exploiting the artificial environment typical of composition classes. They have written to benevolent and understanding instructors but never to faceless and demanding audiences. They have retreated into innocuous topics that they know (or think) have the teacher's sympathy, but they have never had to communicate complex ideas that were totally unfamiliar to a somewhat hostile audience. Above all, they have rarely thought through and defended word choices, sentence structures, and organizational design in terms of the impact each will have upon the intended audience.

Because such frustration is the lot of any professional who writes, we do not allow students to avoid it within the composition classroom; rather, we ask them to reduce that frustration for themselves, first through an analysis of root problems and then through the development of effective responses to those problems. Our assignments are designed to confront students with the range of problems commonly posed by an audience, by a body of information, and by the language itself. Our lectures are intended to make students aware of those problems as a potential source of frustration, to assist students in their detailed analysis, and to help students develop techniques to solve them. Our critical review of student writing points out when students have successfully solved those problems and indicates when problems persist because of student oversight or errors of judgment. The course, then, is shaped by our belief that realistic assignments, pragmatic advice, and functional criticism best prepare students for the situations that will confront them later on as professionals who must write. It is from this perspective that the seven assignments that comprise the course have been developed. (Additional assignments may be obtained by writing the author at the Department of English, University of Texas at El Paso, El Paso, Texas 79968.)

The Seven Assignments

We begin the course by asking students to respond to a letter from Mr. Gallagher (Figure 1), the somewhat bombastic but not unrepresentative Director of Personnel and Training for PROINFO, a firm that, according to our primary scenario, will come to employ most members of the class. From our point of view, this assignment is diagnostic: requiring no research, it could in theory be executed through the employment of skills allegedly mastered within the freshman curriculum. For most students, however, it is a demanding exercise: they must identify and analyze the needs of Mr. Gallagher from the limited data available in his letter; they must collect and evaluate data about themselves, their ambitions, and their expectations; and, above all, they must examine the rhetorical options open to them and select those that effectively organize and express the information needed by Mr. Gallagher.

Because the letter to Mr. Gallagher is a diagnostic exercise, supporting instruction is minimal. In formal lectures we analyze typical university assignments and demonstrate through examples how students often fail to think through the writing problems posed by a given assignment. In discussions we field only those questions about Mr. Gallagher's letter that could be asked about a real firm: e.g., "Does PROINFO hire part-time workers?" Questions about length, form, style, and the personal preferences

PROINFO
123 Technocrat
El Paso, Texas 79968

Dear Applicant:

We are happy to inform you that you are among the 150 applicants whom we are considering for employment with PROINFO. Your records are impressive, and we believe that you would be an asset to our firm.

Because of the large number of applicants and the limited number of positions, however, further screening is necessary. For our purposes, interviews are insufficient: we have found that the skills we seek are best demonstrated in a reflective, rather than a pressured, situation. Thus we request each applicant to submit a written statement that sets forth his or her background, career goals beyond employment with PROINFO, and the type of training he or she expects from our training program. This statement serves two purposes: it gives us an insight into your ability to express in writing a complex idea, and it enables us to evaluate and perhaps modify our training program in the light of the strengths and weaknesses revealed in your written statement.

For your information, let me review the purpose and scope of our firm so that you might better integrate your expectations with ours. PROINFO is a subsidiary of El Paso Enterprises, Inc., a conglomerate with diversified holdings in research and development firms and with contracts in both the public and private sectors. PROINFO has two functions: (1) it is primarily a clearinghouse for professional information that we distribute to research firms, career counselors, employment agencies, colleges, trade publications, and individuals; and (2) it is a source of articulate managers, researchers, and writers for other firms controlled by El Paso Enterprises. Our services are in demand because we carefully research a subject and then supply clients with up-to-date reports that address the problems those clients have posed us. Our products take many forms; thus you, as a specialist in one area, will participate in the preparation of annotated bibliographies, in detailed reviews of current literature in your field, in writing reports on current research, and in constructing other publications unique to your area of specialization. Further, you will be expected on occasion to give oral briefings and to prepare comprehensive packets of information related to your specialty. As you might surmise, our task is not an easy one: we serve an intelligent clientele--experienced managers, college-trained professionals, educators--who need hard, detailed information in a clear and understandable format. Thus we orient our training program to that end: we respect your professional skills, but we must train you to communicate your information to equally intelligent but uninformed clients. This is not an easy task.

We look forward to reading your integrated analysis of yourself, your specialty, and the training we offer.

Sincerely,

Philip J. Gallagher, Director
Personnel and Training

Figure 1. First assignment for all students (5% of final grade).

of instructors are left unanswered, as they could not reasonably be asked of Mr. Gallagher or his real life counterparts. This procedure induces more than a little frustration, but we intend to help students reduce such frustrations as the course progresses.

Submitted after two class meetings, the letters are evaluated according to the impact they would have upon a personnel director such as Mr. Gallagher. We look first for a mastery of mechanics because this criterion is often used as an initial sorting device by personnel managers. We assume that students have been exposed to the "rules" in freshman composition courses; our intent now is to demonstrate the pragmatic usefulness of these rules. Equal weight is given to the student's selection of information that either supports or expands the data already available to Mr. Gallagher through the company's application form. Finally, all letters ought to contain a carefully constructed proof structure that supports the student's primary assertion: because of background, career goals, and expectations, he or she ought to be employed by PROINFO. Each criticism is made in terms of why a given writing strategy would be unsatisfying to Mr. Gallagher.

Most students do poorly on this assignment, but they are not penalized heavily since the letter determines only five percent of the final grade. Through this assignment, however, we have identified patterns of weakness that will shape subsequent presentations; more to the point, students have become more conscious of their limitations as writers and of the need for more rigorous thought in the process of writing.

By the time students have responded to Mr. Gallagher's letter, they will also have chosen either the research option or the PROINFO scenario. For most students, the primary set of assignments (Figures 2, 4–8) will be an extension of the PROINFO scenario initiated by the letter to Mr. Gallagher. Despite the failure of many to satisfy Mr. Gallagher, students are assumed to have been hired as staff writers by that eternal optimist: Gallagher remains confident that they will be able, from their limited experience in their individual fields of study, to communicate career information of value to high school students and to their own academic peers. Mr. Gallagher will, of course, demand that they do extensive bibliographical research—he is well aware that most undergraduates perceive their eventual careers simplistically —but he is confident that with practice they can come to communicate information of value to PROINFO's markets. As instructors, we hope that our students will respond pragmatically

MEMORANDUM

TO: PROINFO Staff Writers

FROM: Projects Director

We have executed a contract with the Texas Council of
Guidance Counselors (College Division) that obligates us to
prepare a series of annotated bibliographies, to be updated
annually, that will provide career counselors on college
campuses with source materials pertinent to the various
career fields open to college graduates. In addition to
standard bibliographical information, each entry should
contain a brief summary of the kind of information
available in each source and an evaluation of that source's
strengths and weaknesses.

TCGC has specified that these bibliographies may
contain up to fifteen entries and that the sources cited
cover all aspects of a given field, including (but not
limited to) educational requirements, current demand for
employees, national and regional wage scales, advancement
opportunities, and current developments in the field.

Each staff writer is expected to produce a bibliography
that is comprehensive, precise, and compact.

Figure 2. Second assignment for PROINFO employees (10% of final grade).

within this context and in the process add to their store of information about their own future careers.

Advanced students, on the other hand, are justifiably not always enthusiastic about the PROINFO scenario; for many of them, the work is superfluous, irrelevant, or, in some cases, too little too late. These students may elect to undertake individual research projects under the supervision of a professor in another department. These projects may take the form of independent research or of research in partial fulfillment of the requirements for a course. In every case, the outside professor's cooperation is essential, and we have taken special pains to inform departments that require this writing course about its aims and methods. Each participating professor must formally agree to provide the student (and our staff) with the necessary technical assistance, including both formal and informal evaluations of the technical content of the student's papers. After we have obtained that consent, we permit students to complete a set of assignments that parallel those in the PROINFO scenario but directly and progressively support their individual research projects.

The second assignment for students who remain within the PROINFO scenario (Figure 2) involves one of the services rendered by the firm and its writers; for students engaged in independent research, the second assignment (Figure 3) comes from the participating professors and marks the formal initiation of their research projects. Both PROINFO and the research directors ask students to gather and evaluate information upon which subsequent work will be based. Both assignments require students to identify the problems inherent in that process and to develop solutions for them, and both assignments prepare students to respond effectively later on as professionals when demands for formal or informal bibliographies are made. Above all, both assignments set for students an audience that they can reasonably be expected to serve. Indeed, counselors on this campus have requested copies of our students' better efforts, while research directors, plagued by student procrastination, have come to require annotated bibliographies for other research projects under their supervision. These immediate applications, when pointed out to students, enhance their involvement in the assignment and provide a credible context for our evaluations of the final products.

The third assignment reflects the fact that the young professional is often asked to review current literature in a field

MEMORANDUM

TO: Student Concerned

FROM: Professor _____, Department of _____

  You are asked to provide the technical writing
instructor and me--and any other interested parties--with
an annotated bibliography of the sources upon which your
research will depend. In addition to standard bibliographical
information, each entry should contain a brief summary of
the kind of information available in each source and an
evaluation of that source's strengths and weaknesses.
While this collection cannot be exhaustive, it should enable
readers to evaluate the soundness of your preliminary
research.

Figure 3. Second assignment for researchers (10% of final grade).

for, if nothing else, the convenience of superiors. The request from PROINFO (Figure 4) and a parallel request from the research directors build upon the material collected earlier for the annotated bibliographies but address the different needs of different audiences: PROINFO is looking to the writer's peers as a potential market, while the research directors intend that the review of literature serve not only themselves but professional colleagues as well. Student writers, then, face two problems: that posed by the information on the subject, information that must be collected, understood, condensed, and structured in a way that makes clear areas of agreement and disagreement, and that posed by an audience who may or may not have preconceived ideas on the subject. Students have the resources to solve both problems: they are (or can become) experts on the problem selected by virtue of focused and extensive reading, and they should be able to see that scrupulous fairness is demanded by their audiences. Students may fail to master the material or to present it effectively, but they have once again confronted credible and demanding audiences very much like those they will face as practicing professionals.

The fourth and fifth assignments focus on descriptions of processes. These assignments de-emphasize the writing of instructions in favor of what can be called operational descriptions (though we discuss and encourage the development, in draft, of the former as an initial check upon the writer's understanding of the process). An examination of professional writings in a variety of settings suggests that most professionals write descriptions of what was done in order to convince the reader that the results of the process under discussion are valid, rather than merely to instruct the reader in the performance of the process. For example, the process we follow in our technical writing course will not be adopted by readers of this article unless that process is demonstrated through our description to be a valid method of instruction. For that reason, the credible description of process receives major emphasis in our course, a description that may include a set of instructions but one that cannot be limited to instructions if it is to be accepted by the reader.

In the fourth assignment, PROINFO writers confront a somewhat skeptical audience with a fully documented description of a procedure or activity typical of a given career field (Figure 5); researchers, in a parallel assignment, submit their experimental procedures for careful professional scrutiny. Because operational

MEMORANDUM

TO: PROINFO Staff Writers

FROM: Projects Director

    The information explosion in all fields has created a
growing isolation among specialists at all levels: as
specialists learn more about their own narrow fields, they
lose command of current topics in broader professional areas.
Our Products Division sees in such specialists a potential
market of some magnitude, and it has directed us to begin a
pilot program to test this market.

    The Products Division envisions a series of publications
that would provide specialists at every level with an overview of current research, recent advances, and controversial
topics prominent in other specialties. This overview would
supply specialists with two types of information: what is
going on in their fields outside of their own specialties
and what debates, problems, and theories remain to be
explored for more definitive results.

    The Products Division asks that I direct each of you
to prepare a thorough and fully documented review of current
literature pertaining to a very specific aspect of your
interest area in support of this pilot project. You are
also asked to include an abstract with the project.

    All projects must be cleared with personnel from the
Projects Director's office prior to full-scale commitment.

Figure 4. Third assignment for **PROINFO** employees (15% of final grade).

MEMORANDUM

TO: PROINFO Staff Writers

FROM: Projects Director

    A consortium of companies has encountered a problem
and, in turn, has asked for our assistance in its solution.
The consortium has found that prospective job recruits are
intrigued by factual descriptions of a firm's activities
and developments. They have therefore asked that we prepare
comprehensive descriptions of such activities and developments to be mailed to potential recruits as a first step
in familiarizing them with the kind of work they can expect
to do if they are employed.

    The Products Division feels that a library of detailed
process descriptions would best answer this request. Each
staff writer is therefore directed to prepare a fully
documented description of a procedure or activity typical
of a given career field. Further, a brief bibliography
should be included as a supplement.

    Questions are to be referred to personnel from this
office.

Figure 5. Fourth assignment for PROINFO employees (10% of final grade).

MEMORANDUM

TO: PROINFO Staff Writers

FROM: Projects Director

Customer response to our last project has been over-
whelming and has created a demand for follow-up material.
Many of the potential recruits receiving the initial mailing
have expressed interest in interviews with consortium
recruiters. Typically, these recruiters are personnel
managers who usually do not have a strong technical background and who, in any case, are not involved with current
developments in a field; yet they are expected to talk in
specific terms about the firms they represent to potential
employees. Thus the consortium has asked that we prepare
materials suitable for oral presentation by recruiters as
they discuss the activities of their firms with prospective
employees.

The Products Division strongly recommends that these
materials take the form of operational descriptions of
typical activities and include quantitative data on the
effects, results, and/or impact of such activities. It has
also informed us that its lithographic facilities are now
fully operational and offer a complete range of reproduction
techniques.

Each staff writer is directed to prepare a fully
documented operational description of a suitable activity.
If descriptions prepared earlier are compatible with the
above specifications, they may be reworked. Further
guidance may be obtained from the staff.

Figure 6. Fifth assignment for PROINFO employees (15% of final grade).

descriptions of such processes are inherently complex tasks, we have added one somewhat artificial restriction: students may not use graphic or tabular aids in their presentations unless absolutely necessary, by which we mean situations wherein the instructor and/or the research director cannot put into words certain information that, therefore, must be presented through tables or illustrations. Supported by lectures, textbook, and staff, the student is challenged to produce a "word picture" that will inform the reader about a selected activity or technique.

The fifth assignment (Figure 6) allows students to recognize and avoid the pitfalls into which they may have stumbled in their first attempts to write operational descriptions. Students are now asked to make full use of graphic aids in their presentations; we also specify that these presentations be delivered orally. Finally, we add a new element to the audience: the professor or recruiter interposed between the writer and the ultimate audience. Such two-level audiences are a fact of life for most professionals. The editors of this anthology, for example, initially read this paper from a perspective quite different from that of the instructors who now turn to this book for ideas to use in developing their own courses; yet the judgment of the editors was crucial to the dissemination of our methodology. In setting such an assignment, then, we attempt to refine the students' ability to present credible operational descriptions while affording them opportunities to experiment with the techniques of oral and graphic presentation. Our primary goal, however, is to make students even more conscious of the complex audiences that will face them as professionals.

Although complete in themselves, the preceding assignments support the student's response to the sixth assignment, PROINFO's request for a comprehensive career information packet (Figure 7) or the requests of the research directors for a compilation of research findings. As a check upon the writer's progress toward that product, PROINFO requires a proposal drafted under its supervision (Figure 8); the research directors make a similar request. These proposals are presented orally to the writer's peers and are revised and approved by them before they are submitted to the appropriate supervisors for the final go-ahead. Neither the career packet nor the research report need be exhaustive—no one can do an exhaustive study every time, and for most of our students, this is one of those times—but the writer must in the proposal define a need for the information to be presented

MEMORANDUM

TO: PROINFO Staff Writers

FROM: Projects Director

    The University of Texas at El Paso has contracted us
to prepare, as a pilot project, a series of comprehensive
career information packets for use by departmental advisors,
the placement office, graduating seniors, and other students
enrolled or planning to enroll in professional or preprofes-
sional curricula.

    It is intended that each packet present both an overview
of the field and a detailed analysis of one subarea; it
should include a sampling of current techniques, research,
and controversies. The overview should provide the reader
with information about employer expectations, wages,
advancement opportunities, and other background. All facts
must be fully documented to aid further investigations by
readers. An information abstract must be prepared for our
records.

    You are reminded that total objectivity is paramount
and that conclusions should be drawn only as warranted by
the facts put forth in your presentation. You are also
reminded that the packet must be an organized whole, despite
the diversity of its contents. You are encouraged to work
with materials already collected, but additional information
will inevitably be needed, as will a carefully thought-out
and appropriate format.

Figure 7. Sixth assignment for **PROINFO** employees (35% of final grade).

MEMORANDUM

TO: PROINFO Staff Writers

FROM: Projects Director

   Preliminary work on the University of Texas at El
Paso project has reached the point where this office feels
that an interchange of ideas is appropriate. Thus several
weekly workshops have been scheduled for the presentation
and critique of preliminary proposals for career information
packets.

   Each writer is directed to prepare and deliver orally
a five-minute description of the direction his or her
research is taking and the final product that will result
from that research. It is imperative that each writer
justify the use of PROINFO time and resources, through that
description, as management auditors are again among us.

   In addition, each staff writer will participate in
these workshops through brief, written critiques of the
presentations made by five colleagues; each critique should
point out the strengths and weaknesses of a proposal and
will be forwarded to the writer making the presentation.

   Please note that while a presentation outlines a
tentative approach to the subject, it should nevertheless
be carefully worked out, information dense, and clearly
written. This office will review the revised versions of
these proposals within a week after their submission.

Figure 8. Seventh assignment for **PROINFO** employees (10% of final grade).

and demonstrate that this need is answered by the information presented. Each work submitted to PROINFO or to a research director is judged on its own merits, which stem from the effectiveness of the work as a response to a clearly defined need on the part of an equally well-defined audience. When students can formulate such definitions and shape their writing accordingly, they have taken a significant step toward becoming effective writers.

Classroom Tactics

Each of the seven assignments in this course challenges students to make informed decisions about audience, information, and writing techniques; the instruction that we offer in support of those assignments does not prescribe those decisions but rather describes ways in which they can be effectively made. In our lectures we outline a variety of approaches that students may use in the analysis of an audience; we examine various ways in which they might collect and assemble information needed by an audience; and we attempt to make them aware of the range of rhetorical options to be considered in presenting the information they have assembled. In addition, our lectures place each assignment within a variety of professional contexts, describing various formal and informal analogues, so that students can better appreciate the potential applications of the skills they are exercising and developing during the course.

The lectures in support of the second assignment—the annotated bibliographies—are typical. Because the student's reaction to a bibliography has been shaped by the almost totally formal requirements of freshman composition courses, we begin by discussing the uses to which bibliographies can be put. In so doing, we help students identify the range of resources available to them in the execution of this project. As they begin to see that they and their colleagues are easily accessible, walking bibliographies and that the apparatuses employed by bibliographical works provide convenient access to the information that they need, they come to understand more about their own purposes in preparing bibliographies for the use of others.

But data collection is only a part of their work: computer bibliographies notwithstanding, most bibliographers reduce their collections to those resources most appropriate to a given audience and its needs. We discuss in lectures and small groups how this might be done, paying special attention to judgments about the relative quality, availability, usefulness, and timeliness of each work. Inasmuch as the annotations required of students are in part evaluative, this portion of the lectures encourages them to articulate how each potential entry meets or does not meet the criteria they are establishing.

Having been advised on the collection and preliminary evaluation of resources, students are ready to turn to the presentation of information. Recognizing that there are as many style sheets as there are professions, we ask each student to identify the one most commonly used in his or her field and to master its intricacies by using it in all work done in the course; we look, in our evaluations of a student's work, for a growing command of the appropriate style sheet. At the same time, we make an effort to explain the functions of such forms, demonstrating how they facilitate the audience's use of the information presented through them. Because there is considerable resistance against the use of these forms, our explanations must be a persistent and important part of subsequent lectures.

More to the point, however, we discuss various techniques of presentation that are available for use by writers in the professions. Our intent here is to show how such techniques can be "invented" by the writer and then evaluated for their usefulness and appropriateness in a given context. We specify no one structure, sequence, or format for any project; instead, we ask students to identify and employ those that effectively set forth the information needed by the audience. We attempt to demonstrate that what is clearly required in one situation may be clearly inappropriate in another, and that the writer is accountable for these decisions.

Finally, the consequences of such decision-making are brought home through the examination of student writing from previous semesters. We concentrate first on a writer's failures and their consequences for the reader: students, now functioning as the audience, can often deduce what the writer intended to do, but they can also appreciate, from their own difficulties with the text, the problems that the writers failed to solve. From this critique we move to a discussion of how the writer might better have served the reader. These critiques prepare students for the objective analysis of their own drafts prior to final typing and submission.

We also use small group work to encourage the critical analysis of work in draft. Here students, having prepared rough drafts, are called upon to analyze the effectiveness of one another's efforts. Experience in group work is especially important in the technical writing classroom because cooperative efforts, almost inevitable in the professional world, have been implicitly discouraged by the operations of most college classes. This cooperation grows slowly, and it must be carefully supervised to ensure productivity; but its value is inestimable because it allows students to test and validate for themselves the information presented in lectures as they work towards the improvement of their own drafts.

The annotated bibliographies, like the other assignments, are not difficult to evaluate: those for the research projects are aimed, in part, at the staff, though they must primarily serve the research directors; the bibliographies written for PROINFO are immediately appreciated by the instructor who, like the career counselors, can make use of them in counseling future students. Again, we pay careful attention to the mastery of mechanics since errors interfere with the presentation of information and often influence the reader unfavorably. We watch for completeness: parallel assignments from students working for PROINFO provide a standard of completeness for career bibliographies, while the evaluations of the research directors provide a functional measure of completeness for research bibliographies. Annotations are evaluated for clarity and for the assistance they provide readers; criticism of the annotations is expressed through questions that the instructor can reasonably ask but that were left unanswered by those annotations. The apparatuses employed by each writer are examined first for consistency and then for effectiveness: if the instructor can identify a structure that has demonstrably greater usefulness for the audience, then the student's selection has been faulty. We believe it to be vitally important that each criticism be functional, that it be understood in terms of the damage done by the item criticized. With each criticism, however, the instructor should be able to suggest an alternative that is demonstrably more effective. In a course such as this, the assignments, the lectures, and the evaluations must be pragmatic if they are to enhance the student's writing in the future.

On the whole, this is a difficult course. The workload is heavy, especially in the first half of the semester during which the second, third, fourth, and fifth assignments fall due at two-week intervals; but the sequence is designed to concentrate most of the bibliographical research in the early part of the semester instead of at the end, the time when papers and projects in other courses tend to fall due. The memos defining each assignment require interpretation: they are not models of style, but they do set the tasks in generalizations typical of many supervisors. The lectures are not prescriptive, but neither is the assistance or instruction that young professionals receive on the job; students must learn to take advice without relying slavishly on it. Finally, the grading is rigorous, if not harsh: fewer than twenty percent of the students earn passing grades on the first assignment, for example, but only that percentage of the students have written letters that would not elicit an even harsher judgment—"Don't call us, we'll call you!"—from a personnel manager looking for competent young professionals. We represent the last opportunity for students to make mistakes and to learn from them without paying the high price of failure exacted in the professional world, and we would be doing students a disservice if we did not exercise and inform their abilities as writers in a way that reflects as closely as possible their employment in the professional world. This course attempts to make the most of this instructional opportunity and to serve the very real, if sometimes unappreciated, needs of our students. We believe that it succeeds.

2 The Case Method: Bridging the Gap between Engineering Student and Professional

Ben F. Barton and Marthalee S. Barton
The University of Michigan

To bridge the gap between the classroom and the world of the professional is one objective of technical communication courses. For students with work experience this objective poses few problems. For others, however, the problems, audiences, and communication needs of professionals are remote. A case method that encourages students to analyze open-ended problems, to adapt to audience needs, and to practice team writing is one way to reach these students.

As a means of preparing students for professional practice, the case method has a long history. Originating in the fields of law and medicine, it has for some time been a staple pedagogical tool in modern business.[1] The last decade has witnessed its successful extension to other fields, such as engineering.[2] For teachers of technical communication trying to bridge the gap between student and professional, the method would appear to hold promise; yet, as a review of the literature shows, that promise is largely unfulfilled.[3]

Here we examine the case method as a tool for addressing some of the problems encountered in teaching a course in technical communication. We turn first to a description of the course and some of the associated problems. We then describe the basic features of the particular case method used—the choice of case problem, the resources made available to students, and the communication tasks and exercises assigned. Finally, we describe advantages and disadvantages of the method and potentially useful variants.

The Course and Its Problems

We teach a senior-level, multisectioned course in technical and professional communication in the College of Engineering of the

University of Michigan. The course objective is to train engineering students with a wide variety of specializations to write professional reports that are instrumentally useful for diverse audiences in organizations.[4] Course assignments generate technical communications in which problems are formulated and solutions advocated for such audiences.

For a few of our students, these assignments pose no special difficulty. These students have already worked professionally and need only select from their files reports that they can adapt for purposes of the course. In contrast, most of our students have not had professional experience and have not written reports that have instrumental value in actual organizations. They are required, therefore, to devise reports for imagined audiences, and typically they have difficulty doing so. Difficulties arise in handling organization, in treating technical issues, or, because of their interrelatedness, in both.[5]

Such difficulties arise largely because these students misconceive the role of the professional engineer—a misconception that is reinforced when they cling to the role they are accustomed to playing, that of the engineering student. For there are radical differences between the two roles—differences that lie largely in the nature of the audience, purpose, and problems addressed by students and professionals.[6] That is, students write for a single, authoritative audience—the professor—to demonstrate a mastery of subject matter. They tend to treat problems that are tutorial in nature—that is, preformulated and formal, or context-impoverished problems with predetermined solutions. Professionals, on the other hand, write for multiple, diverse audiences—some more knowledgeable than they, some less. Moreover, they write largely for instrumental rather than for informative purposes; that is, their primary goal is to accomplish something for the organization to which they belong. Unlike students, they tend to treat problems that are open-ended and ill-defined, occur in a rich context, and are amenable only to provisional solution.

But students misconceive more than the professional role in its relation to the communication process; they also misconstrue the general nature of technical communication itself. Specifically, their view of the technical communication of engineers is overly narrow: Technical communication is thought to take place *after* a technical problem has been solved, to be an "art" concerned largely with matters of arrangement and style. In short, students are unaware that communication, like engineering, is a process—a mediating between perceived need and desired effect. Thus, they

Table of Contents

Figure 1. Casebook contents.

are largely unaware of the transactional character of technical communication. These misconceptions are, unfortunately, reinforced when students are asked to report on engineering problems that have already been formulated and solved.

The Case Method as Solution

Our awareness of these misconceptions led us to seek an approach that would exploit the fact that the arts of engineering and communication are, in fact, inseparable activities—activities that share such critical features as a problem-solving methodology and a common goal of product instrumentality. Specifically, we sought to confront inexperienced students with a set of carefully metered demands for defining, solving, and reporting an authentic engineering problem within an organizational context. This objective led us to develop the case method discussed below. Our discussion focuses on one variant of the case method—the use of the case by students in a section of the technical and professional communication course described earlier.

Three criteria underlay our choice of case problem, namely, that the problem be "real," of general interest, and of circumscribed difficulty. Most important, we sought an authentic problem—one that would represent problems typically encountered in practice by entry-level engineers. Second, we wanted a problem that could be handled without a deep understanding of concepts peculiar to any one engineering specialization, a problem that would permit a focus on the structural paradigm underlying all engineering specializations as well as rhetoric itself, i.e., the problem-solving methodology.[7] Third, we desired a problem that could be treated adequately in a one-term, technical communication course and that would not divert students from rhetorical issues. While many suitable problems exist, our initial case involves the choice of fire-warning systems for subdivision homes being planned by a hypothetical construction company.

Case Resources

The following resources are available to students working on the case problem.

The casebook. The nature of the materials provided in the casebook is suggested by its contents (Figure 1). Section 1, the introduction to the casebook, orients students to the case method.

Section 2, Description of the Bellevue Construction Company, describes the nature of the firm's business and its general method of operation. An organization chart of the firm is provided, and the responsibilities of its various organizational components are clarified. Further, Section 2 positions casebook users within the organization and gives them specific roles within the project. The communication demands that arise in carrying out the project constitute the rationale for the sequence of five increasingly complex communication tasks set out in Section 3. (A representative task of intermediate complexity is presented later in this paper.) Other materials useful in completing the assignments are provided in the casebook appendices: selected bibliographies of nontechnical and technical references, selected lists of organizations, agencies, and manufacturers, worksheets for characterizing smoke and heat detectors, and selected articles and brochures on fire-warning equipment.

Class handouts. Students are provided periodically with supplemental class handouts. Some of these may eventually be placed on library reserve or incorporated into a later version of the casebook. Occasionally, these materials update technical information already in the hands of students.[8] More often, handouts take the form of exercise sheets used to clarify rhetorical issues. (Two sample exercises are presented later in this paper.)

Materials on library reserve. Other resource materials—not included in the casebook because of their bulk or proprietary nature, for example—are placed on library reserve. These materials supplement those in the casebook and include additional articles and reports, codes and standards, manufacturers' brochures and specification sheets.

Other resources. In addition to materials in hand and on library reserve, the general resources of the library are available to students. To facilitate the exploitation of library resources, students attend two orientation lectures early in the term. The first, delivered by a professional librarian, provides a general introduction to methods and tools for information retrieval. The second deals with specific resources relevant to the case problem, for example, how to locate case-related materials such as building codes. The lecture also initiates students into the art of computerized search, using keywords drawn from the case problem. More importantly, perhaps, the reference librarian serves as an important consultant to students throughout the term. Students are also informed about other consultative resources available in the community, for example, manufacturers' representatives, building inspectors, and personnel with local building firms.[9]

These resources are more than adequate to complete the case assignment. Using them, students gain direct experience in information accessing; at the same time, the case method ensures that demands for accessing needed information are not excessive. Armed with information largely assembled by the instructor, students are freed to focus on the assimilation, evaluation, and arrangement of the information for given rhetorical purposes. These rhetorical purposes are delineated in the following section.

Case Tasks and Class Exercises

The casebook calls for the completion of five increasingly complex communication tasks. The first is the preparation of a relatively simple letter of inquiry requesting needed information about a specific smoke detector; the last calls for a comprehensive, formal report. With the exception of the final report, students are given optional topics for each communication task. A task of intermediate complexity is given below. The most convincing student response developed three main reasons for using smoke rather than heat detectors: the exclusive use of heat detectors violates codes; heat detectors are ineffective, e.g., slow, in responding to the common "smoky" fire; an effective system incorporating only heat detectors would, in fact, be too costly.

> L. L. Nehru, Head of Purchasing, has forwarded (in Memorandum W.23, dated 17 September 1980 and addressed to Y. S. Amed, Head of Engineering) the cost quotations on selected smoke-detector units requested by Mr. Amed in Memorandum W.7, dated 9 September 1980. In his memorandum, Mr. Nehru notes the widespread availability of relatively inexpensive *heat* detectors. Further, the memorandum suggests that as an economy measure heat detection be used exclusively in the fire-warning systems of Woodview Subdivision. Mr. Amed has delegated you to respond to this memorandum.

Complementing the formal case tasks is a series of exercises— some done individually outside of class, some done in class workshops. These exercises introduce, illustrate, and provide practice in specific rhetorical skills. An exercise used in a class workshop early in the term follows.[10] It is designed to familiarize students with the hypothetical organization of the case company and to give them practice in identifying audiences.

You are an entry-level staff engineer working for Y. S. Amed, Head of Engineering at the Bellevue Construction Company. (See pages 5 to 8 of the casebook for a chart of the company organization and a description of the responsibilities of its various organizational components.) Your current assignment involves the design of fire-warning systems for 140 homes to be built next year in Woodview Subdivision.

At a recent staff meeting, B. N. O'Neal, Head of Marketing, reported that sales personnel have noted widespread public debate about the radiation hazard posted by ionization smoke detectors. He suggested, therefore, that such detectors not be used in Woodview Subdivision. After staff discussion, J. L. Bonet, President of Bellevue, requested a formal opinion from Engineering (in follow-up Memorandum W.36, dated 30 September 1980). Mr. Amed has delegated you to prepare a responding memorandum.

Identify (names and roles) and briefly explain the audiences for your memorandum: *primary audiences* (those who make decisions or act on the basis of the information a report contains), *secondary audiences* (those who are affected by the decisions and actions), and *immediate audiences* (those who route the report or transmit the information a report contains).

The exercise shown below gives students experience with heuristic procedures for the retrieval of information, sensitizes them to the conventions of referencing, and familiarizes them with the recent literature on the case problem. (It also provides a convenient means for updating the instructor's case bibliography.)

You have heard two lectures by a reference librarian in which heuristics for information retrieval were described. Your assignment is to use these heuristics to identify three information sources related to the case topic and dated after December 1, 1979—the approximate cut-off date for the bibliography on pages 23 to 29 of the casebook. Complete and submit the attached flow diagram, thereby detailing the steps taken in locating one of your sources. Describe, in the space provided, special problems you encountered.

In addition, prepare and submit two sets of bibliographic entries for the three sources. The format of the first should conform to standard practice in the field of your engineering specialization. (Electrical engineers, for example, should follow the format delineated in the appropriate issue of the *Proceedings of the IEEE*.) If necessary, consult members of the faculty in your engineering specialization to determine the appropriate standard in your field. In the second, adopt the format used at the Bellevue Construction Company and shown in the casebook bibliography. Clearly identify the standard used to prepare each set.

An Evaluation of the Case Method

The case method permits a carefully controlled simulation of professional activity in a one-term, technical communication course. In particular, students are provided the basic information to fulfill, in a professional manner, a carefully circumscribed assignment. An organizational context similar to those typically given to entry-level engineers is provided for the assignment. Students participate in all phases of the professional communication activity, from problem formulation and information accessing to editing the final report. At the same time, the assignment is feasible because demands for certain time-consuming phases of professional activity—for example, information accessing —are moderated. Though students are encouraged to exploit information from a wide variety of sources, and are given help in doing so, much of the needed information is conveniently available in the case materials assembled by the instructor.

A case provides a unifying theme for a course in technical communication because the case assignment is itself coherent. Moreover, the case can be closely tied to the formal instruction in the course through a sequence of communication tasks and complementing exercises. The unifying effect of the case easily compensates for the initial investment that students must make to master the technical matter on which the case is based.

One frequently cited advantage of the case method is its capacity to heighten the perceived relevance of course work and thus to increase student motivation. In the words of G. H. Flammer, "One of the biggest benefits realized from case studies is student perception of the reality of problems and the relevance to his or her imminent professional practice. Perceived relevance is a strong motivator."[11] Furthermore, students perceive formal instruction in a communication course as more relevant when it is tied to the subject matter of a common case assignment. For students in a traditional communication course, the technical matter used to illustrate a given rhetorical principle is often of little interest— indeed, it is often an impediment to learning. On the other hand, students working on a case assignment find the materials used to illustrate rhetorical principles of interest both technically and rhetorically.

Another advantage of the case method is its adaptability to other pedagogical methods for introducing students to professional practices. In fact, the case method may serve to enhance the effectiveness of those methods. Consider the interview visit—a tactic that introduces questions of how, why, and for whom reports are written in organizations. The key ingredient is, again, perceived relevance, and the case assignment adds just that ingredient. A visit to, say, a local architectural engineering firm has immediate instrumental value for the student who must write a report specifying the details of placement and installation costs of smoke detectors.[12] The case method is equally adaptable to role playing—another pedagogical tool that bridges the gap between student and professional. Class members, for example, assume the roles of various professionals in the Bellevue Construction Company during question-and-answer periods following technical briefings by students.

The case method also provides opportunities for students to gain experience in team writing—a common mode of professional communication. Small groups—typically three to five students—may collaborate in writing the final, formal report. Such a collaborative effort might be required by the instructor but has to date been undertaken in our classes only by students who volunteered and who, we should add, were strongly disposed to accept the challenge. Professionals, however, often must collaborate in the production of reports, yet opportunities to master the special skills needed to participate in team writing occur infrequently in the traditional curriculum. The opportunity to develop skills in team writing has, in fact, been cited by several students as the single most worthwhile experience gained from work on the case assignment.

Perhaps the most gratifying advantage is that the case method directly addresses student misconceptions about the nature of the communication activity of professionals. Specifically, the case method breaks down the false dichotomy between the technical and the rhetorical. Moreover, the case may provide the most viable means available for effecting the cooperation between technical and rhetoric staffs that is so widely advocated in the literature.[13] That is, collaboratively-generated case materials can be used effectively in technical communication courses without the sustained involvement of the technical faculty.

Two disadvantages of the case method arise when, as in the example discussed here, a case problem is assigned to all students in a class. Although numerous benefits accrue from sharing a substantial body of knowledge, there is an attendant cost because the class cannot serve as the diverse audience typically addressed by a professional report. Students working on a common case assignment understandably find it difficult to adopt the perspective of readers largely unfamiliar with the subject—an important aspect of the diverse audiences in organizations. In other respects, however, students familiar with the subject matter of a report are far more effective critics. A second and perhaps more serious disadvantage is that students cannot choose their own communication topics, as some might well prefer. Admittedly, they may choose within a limited range of topics, but they cannot usually draw directly on the subject matter of their particular engineering specializations. This limitation can be reduced by developing cases for specific disciplines, and we are, in fact, now assembling materials for several such cases. The problem of over-restriction can also be alleviated by adopting some of the methods described below.

While we have focused on a specific variant of the case method, one of the chief advantages of the method is, in fact, its flexibility. In another variant, for example, the case method was followed —in an otherwise conventionally taught class—by a group of students who desired experience in team writing. In another conventionally taught class, the method was used to introduce engineering methodology to one nonengineering student.

Clearly, many other variants of the basic method are possible. One particularly promising strategy is to introduce the case at the underclass level, an especially appropriate strategy because underclass students almost assuredly do not have professional or design/project experience. Moreover, they have little discipline-specific knowledge on which to draw for reports. Thus the case method introduces such students to the problem-solving methodology underlying the arts of both engineering and communication.

In other variants, formal assignments might not be limited to a given case problem. One might, for example, use a case as the basis for several assignments early in the term; students could later report on problems of their own choice. Moreover, the case may be a valuable instructional resource even if it is never used as the basis for formal assignments. That is, the use of case materials may be justified if only to provide a unified set of examples—including model reports by past students—that helps students understand the various and complex features of authentic communication problems. Such is, in fact, the fate we envision for old casebooks.

Notes

1. For a brief history of the case method, see D. Little, "The Case: Milieu and Method," *The Journal of Business Communication* 8, no. 4 (Summer 1971): 29-35.

2. The February 1977 issue of *Engineering Education* features a series of articles reporting successful uses of the case method in engineering courses. The growing number of cases available indirectly attests to the perceived promise of the method. Two collections of engineering cases have already been published: H. O. Fuchs and R. F. Steidel, eds., *Ten Cases in Engineering Design* (London: Longman, 1973) and K. H. Vesper, *Engineers at Work: A Casebook* (Boston: Houghton Mifflin, 1975). Further, the Engineering Case Library of the American Society for Engineering Education (ASEE) includes several hundred cases ranging over diverse disciplines.

3. R. Hays notes the unfulfilled promise of the case method in his article, "Case Problems Improve Tech Writing Courses and Seminars," *Journal of Technical Writing and Communication* 6, no. 4 (1976): 293-98. Hays then describes an important first step in bringing the case method to technical communication courses. In his approach, the case problem is an exercise which "can be a handout of from one to four single-spaced pages. The first paragraph or two of the handout will tell the students or trainees how to do the assignment. The rest of the handout will be data—statements of facts, quotations, lists of figures, short tables, and citations of opinion . . . randomly listed, stated in fragments, and sprinkled with mechanical errors." Students are asked to organize the data, eliminate irrelevancies, invent missing data, and write a report for submission within a few days.

4. The course was developed by J. C. Mathes and D. W. Stevenson of the University of Michigan, and their book, *Designing Technical Reports: Writing for Audiences in Organizations* (Indianapolis: Bobbs-Merrill, 1976), is the basic text for the course.

5. For an overview of difficulties students encounter when required to simulate an organizational context, see P. R. Klaver, "Writing as Engineers and Writing in Class: Simulation as Solution and Problem," in *Technical and Professional Communication: Teaching in the Two-Year College, Four-Year College, Professional School*, ed. T. M. Sawyer (Ann Arbor, Mich.: Professional Communication Press, 1977), pp. 155-66. For a discussion of the difficulties students encounter when required to formulate and articulate a suitable technical problem, see B. F. Barton and M. S. Barton, "Toward Teaching a New Engineering Professionalism: A Joint Instructional Effort in Technical Design and Communication," in *Technical and Professional Communication*, pp. 119-28.

6. Our discussion of the distinction between student and professional on the basis of audience and purpose draws heavily on chapters 1 and 2, Mathes and Stevenson, *Designing Technical Reports*.

7. For relevant scholarship on the treatment of the rhetorical act as problem solution, see, for example, R. E. Young, A. L. Becker, and K. L. Pike, *Rhetoric: Discovery and Change* (New York: Harcourt Brace & World, 1970), especially chapter 5.

8. These technical handouts are often generated by students and, at times, have unexpected uses. Last term, for example, a student brought in an article that projected a significant advance in smoke-detector technology, the then-impending commercial availability of low-cost, high-performance, integrated circuits for smoke detectors. Though the article described a technology that could not be exploited in the case assignment because of time constraints, it served to illustrate dramatically the provisional character of solutions to engineering problems.

9. Though we have not done so to date, appropriate outside resource persons might well be invited to participate formally in a technical communication course using the case method. We note in passing, however, that numerous enterprising students have interviewed widely among the available consultative resources.

10. The taxonomy of report audiences adopted in this exercise is taken from Mathes and Stevenson, *Designing Technical Reports.* For an extended discussion of the three types of audiences (primary, secondary, and immediate), see pp. 21–22.

11. "The Case Study: Exercise in Simulation," *Engineering Education* 67, no. 5 (February 1977): 372. The role of cases in increasing student motivation is more fully discussed by Flammer in "Applied Motivation—A Missing Role in Teaching," *Engineering Education* 62, no. 6 (March 1972): 519–22. See also H. O. Fuchs, "On Kindling Flames with Cases," *Engineering Education* 64 (March 1974): 412–15. Both Flammer and Fuchs base their claims on experiences with case problems in technical courses; however, their basic point has general applicability.

12. Perceived relevance is clearly the basis for the injunction of J. C. Mathes and D. W. Stevenson that the student "arrange an interview with someone whose discipline and role are similar to those for which the student is preparing." See *Designing Technical Reports: Teacher's Manual* (Indianapolis: Bobbs-Merrill, 1976), p. 15. A similar injunction is offered by J. Halpern, whose experiences with the interview technique are reported in "Interviewing in Business and Industry: An Effective Way to Introduce Purpose and Audience to Students of Technical Writing" in Sawyer's *Technical and Professional Communication*, pp. 139–54. An interview on the case assignment satisfies their injunction, in that a typical visit involves contact with an entry-level engineer. Thus, a case interview has the kind of perceived relevance and general instrumental value sought by the above commentators in addition to the immediate instrumental value noted above. We mention in passing that interviews with professionals in other roles, e.g., managers or executives, may yield further insight into the problems of communicating effectively with diverse audiences.

13. Sawyer's anthology alone contains four articles dealing with a single approach to cooperation between technical and rhetorical faculty—that is, team teaching.

3 Communication Strategy in Professional Writing: Teaching a Rhetorical Case

Linda S. Flower
Carnegie-Mellon University

The communication strategy required by writers in their professional lives is not the same as the paper-writing strategies acquired at school. A technical writing course that focuses on communication strategy, especially through the use of the rhetorical case, provides students with compelling reasons to write, audiences that need to know, and meaningful roles as writers.

Most of us who teach professional writing face a peculiar problem —not only are we trying to figure out how best to teach our subject, we are still trying to define it. For years teachers of technical writing have debated that apparently unanswerable question: What characteristics distinguish technical writing from the kinds of writing traditionally taught in composition classes?[1] It might help to change the question. Instead of examining the product, perhaps we should study the writer, asking instead, "What skills do professionally effective writers—whatever their jobs—need?" The purpose of this discussion, then, is threefold. First, I suggest that the foremost goal of a professional writing course is to teach students to develop a *communication strategy*. Secondly, I compare the communication strategy that professionals need to the paper-writing strategies that students normally acquire in school. Do these two sets of skills match? Are we, in fact, preparing students for the writing they will do in their professional lives? Finally, I look at some of the ways we can teach communication strategy, with special emphasis on the rhetorical case.

Writing in Professional Life

English teachers sometimes treat technical or business writing as a set of quite specialized skills needed by "other people"—by

34

engineers, managers, lawyers, or scientists—people who have "technical things" to say. In fact, professional writing is the writing all of us do after we leave school. The format may indeed vary among professions, but whether we are in a research and development lab, a marketing group, or a university department, our reports, proposals, and memos to colleagues are all professional writing with an important feature in common: unlike school writing, they were generated by an authentic purpose and written to communicate information an audience doesn't know, but should. Writers in the professions write because they need to make something happen.

Effective writers in the professions get proposals funded, convince colleagues to act, enable readers to understand the point and value of their ideas. My hypothesis here is that the critical skill behind effective professional writing is the ability to develop a communication strategy that makes those things happen. We may at first be suspicious of strategy in writing because of its association in classical rhetoric with the art of sophistry and persuasion by any means. However, unlike a debate strategem, which works on an "I win/you lose" premise, communication strategy is a mutual affair between writers and readers. Writers in the professions must transform, reorganize, maybe even reconceptualize information and ideas in order to communicate them to the reader. The act of transformation is the key.

Let me give you an example of how a professional's communication strategy might work. Suppose architect Nancy Brown were asked to talk to a group of city commissioners on the subject of energy-conscious construction. Thinking about the talk, she might quite naturally begin with her own knowledge of architecture organized under such categories as methods, materials, styles, and functions. We can imagine this information arranged in categories and stored in a giant loose-leaf encyclopedia of the mind. In giving her talk to the city commissioners, she could simply work her way through this established network of ideas, flipping through these mental pages for relevant entries on energy. If, on the other hand, Ms. Brown really wanted to move the commissioners to action, she might develop a communication strategy that would lead her to reorganize her information into new categories that she and the audience would share: Major Sources of Energy Loss in Buildings, Practical Measures to Conserve Energy Now, Conservation Measures for the Future. That communication strategy would take into account the conceptual framework of the commissioners, a framework quite different from that of the architect, and their need to use this information to act.

In essence, then, when a communication strategy works, the writer meets his or her goals by meeting the needs of the reader. Carrying out a strategy may be as simple as translating jargon or reorganizing a paper, or it may be as demanding as reconceptualizing one's ideas to fit a framework that can be shared by the reader. In either case, and this is what I want to emphasize, a communication strategy is a self-conscious attempt on the part of the writer to get through to the reader—to communicate, not merely to express.

It may seem that a communication strategy should come as naturally as breathing. In face-to-face communication it often does because immediate feedback lets us test our ongoing message and adapt it to the listener. But writers work in private, and strategy-making is hard work. As a result, many writers fall back on merely expressing what they know and letting the reader do the work. An example is the sales representative's report that relies on a narrative or journal format. The reader, perhaps a marketing manager, may depend on that report to predict trends and spot trouble: Should we market more insulation this year? Do consumers need further installation information? The sales representative who has not gauged that reader's needs may merely chronicle how he or she spent the day, leaving the manager to ferret out the pertinent information.

Writing in School

My hypothesis to this point has been that the critical skill required in professional writing is the ability to integrate the writer's *purpose* and *knowledge* with the reader's *need*. If this is true, then the next question we should ask is this: Are the same set of skills required to be a successful writer of classroom compositions and typical college papers? How do the strategies we develop as students stack up against those we need as adults?

We could look at this question in terms of priorities. The priorities for many school assignments are placed on correctness and acceptable form or on having good ideas, no matter how they are expressed. In either case, a communication strategy is not particularly necessary. Students are writing to an expert on their subject, an expert who reads to evaluate their message, not to use it. In professional life, however, the priorities are quite different. Form and style matter most only when they are violated; they operate at the level of a minimum standard, as does correctness itself. They can matter, but they are normally secondary to the reader's need to use what the writer knows. In fact, even a "good idea" matters only if the writer can communicate it to someone else. Professional writers must often do more than write clear or correct prose; they must be communication strategists. Yet school gives them little chance to test and thereby develop mature strategies for communicating with other people.

The gap between school and professional writing is widened by a further irony. Not only do students fail to acquire strategies for dealing with an audience, but many of the paper-writing tactics they do learn become downright liabilities when they go to work. Let me mention three that are familiar and widely used: the state-and-elaborate strategy, the textbook strategy, and the what-the-teacher-wants strategy. Each of these cuts the large problem of communicating down to the simpler problem of producing a paper. By a selective emphasis on only one element in the communication triangle of writer, reader, and subject, each of these strategies effectively reduces the communication problem to a level of minimal constraint.

According to students I've known at four universities, the most popular strategy for writing papers in college is to state and then to elaborate.[2] State-and-elaborate is essentially an expressive strategy. The structure of the message reflects the structure of the writer's thought; its purpose is to demonstrate what the student thinks. The audience, and sometimes the subject itself, is a minimal constraint. The method is relatively simple. First you get an idea. (This part is usually left to inspiration.) Then, whether you jot down an outline or write as thoughts come, you elaborate on your idea until you run out of information or reach the page limit. If you use acceptable form and style, this strategy often works in school.

An alternative to self-expression, or the burden of having one's own ideas, is the textbook strategy. This tactic allows the student to print out, like a computer, a restatement of what he or she has learned. The structure of an outside body of knowledge dictates the structure of the paper. Like the talk on energy our architect could have given, this strategy avoids reconceputalizing, perhaps even thinking about the assigned subject at all. The audience's need is a minimal concern. In addition, this strategy ignores the writer's goals and thoughts as well. The information speaks for itself, or so the student assumes.

The third strategy for writing in the classroom may be more prevalent than we wish to recognize: writing what one thinks the teacher wants to hear. The effect of this strategy is to minimize the importance of writer and information and to focus primarily on the reader: to tell the reader what he or she already thinks and therefore presumably wants to hear.

Each of these writing strategies is ill-adapted to the larger problem of communicating. They allow the writer to lop off the "extra" demand of an audience with needs of its own and to deny the existence of a valid purpose for the writing task. These are luxuries rarely afforded to professionals, who write in the context of an authentic rhetorical situation in which writer, reader, and subject all place powerful demands on the act of writing.

The problem then is this: students tend to develop strategies for writing that are clearly based on the specialized demands of the school assignment. When they leave school, the rhetorical situation changes dramatically, but those ingrained paper-writing strategies may not change at all. When this happens, a writer's strategies are not just inadequate, they are a liability. They lead the writer to reduce a complex communication problem into the familiar but artificial task of writing a paper.

Developing Courses in Professional Writing

The difference between the demands of school and professional writing is a problem writing courses must overcome. If the typical composition course is not adapted to teaching professional writing skills, how do we design courses that are?

One option is simply to borrow the old clothes of freshman composition. We could treat business or technical writing as a special genre and focus instruction on questions of form and style: the form of the business letter, the parts of the technical report, the evils of jargon, the sin of the passive. However, like the paper-writing strategies we have just discussed, this tactic would fail us because it ignores the demands of authentic and complex rhetorical situations. Instead, we need to devise assignments that offer realistic communication problems for students to solve. We need to throw students into a full-bodied version of what Lloyd Bitzer calls the "rhetorical situation," a situation where there is a genuine need to write, a demanding audience, and realistic constraints.[3] Student writers need to meet situations that require them to take a "rhetorical stance."[4]

Teaching assignments that take this plunge range from the highly realistic and therefore relatively uncontrolled project to the narrowly focused rhetorical case. Internships and project courses allow students to take on assignments within an organization such as a business, a civic enterprise, or a university.[5] There they encounter the unexpected and deal with the political and personal forces that impinge on professionals when they write. Simulations and games, on the other hand, select and limit these forces, bringing them within the scope of several class sessions during which students play various roles within a simulated business and face a set of predictable communication problems.[6] Like project courses, simulations serve a unique affective function —student writers experience the results of their actions. As we move across this spectrum from the relatively uncontrolled project to the more restricted case, we lose in realistic complexity but gain in focus and teachability. That is, the more controlled the assignment, the more readily students are able to transfer specific textbook techniques to practice.

Somewhere between project courses and case studies fall problem-solving assignments that ask students to identify a practical problem encountered at school or work and to write a consulting report that helps solve the problem. The instructor can suggest campus-related problems (scheduling at the counseling center, how to generate material for the campus paper) or leave the assignment quite open:

> You are a free-lance consultant. Your assignment is to analyze a problem your client has encountered. Research the situation, define the critical issues, and write a report that helps your client solve the problem.

It is possible, of course, to tailor the assignment to require specific skills you wish to teach. For example, you can stipulate that the report be designed for both management and user audiences. But the heart of the assignment is analyzing a problem arising out of an authentic, complex situation and writing a document that actually works for its readers.

One student, for example, drew on a summer job experience in an auto repair shop. The mechanics in the shop and the personnel in the parts department were at loggerheads—each group needed information that the other wasn't prepared to give. The student not only had to solve the practical problem (he designed a new system of order slips and progress reports), but he also had to solve the rhetorical problem of explaining the system and persuading the mechanics and parts people to adopt it.

Problem-solving reports throw students onto their own resources; the teacher is an advisor in an experience that may take several weeks to complete. By contrast, a rhetorical case may require only a single class period to complete and generally focuses on only one or two central points. In losing breadth and complexity, however, the case gains the power of a highly goal-directed activity.

Designing and Using the Rhetorical Case: Memo to the Dean

Rhetorical cases are specifically focused, goal-directed teaching tools. And yet, their ultimate goal is to help students test and develop communication strategies that they will be able to use outside the classroom. In designing a case to meet these ends, I try to include four elements:

1. A realistic problem or exigency—a compelling reason to write.
2. A demanding audience that needs to know or act—not a teacher or an evaluating expert.
3. A meaningful role for the writer—not a student role—that includes a clear understanding of the goal to be achieved by writing.
4. A body of information presented in the unsifted, temporal form in which writers in the professions usually uncover it. The case should give facts; the writer's job is to generate concepts and create a structure.

The rhetorical case differs from a normal business case in that it focuses squarely on a rhetorical problem, not a management problem. The special skill required is rhetorical strategy rather than the ability to make decisions or an understanding of organizational psychology. In the case to be developed here, Memo to the Dean, the focus is on Rogerian argument, a mode of persuasion in which the writer tries to argue without polarizing the issues and works to create a ground of mutual agreement.[7] Although within the context of the course this case follows a unit on argument, it is assigned without instructions to use a particular kind of argument. Its teaching power is this: it is not presented as an exercise in Rogerian argument but as a problem that the writer must solve. As it turns out, a form of Rogerian argument is the best solution.

The instructional goals of the assignment are two: to help students recognize cues in a rhetorical situation that call for a particular communication strategy and to encourage students to treat writing as a problem to be solved through the development of a communication strategy that is of mutual benefit to writer and reader.

This case was designed for both sophomore and graduate business students. Older students naturally bring more savvy to the problem; however, the assignment encourages younger students to tap native rhetorical skills they often exclude from school writing. It confronts students with two specific problems. First, they must fit their "good ideas" into the Dean's frame of reference and solve a problem that both they and the Dean recognize as important. Second, they must overcome the Dean's bias for another plan. To handle this case, their first task is that hardy perennial of communication—transforming information instead of merely expressing it. The second problem, the Dean's bias, is deliberately planted in the case to call for Rogerian argument. In reading the case assignment below, notice how it presents unsifted information in much the way that information is encountered in professional life.

> You are on the faculty of a small, aggressive graduate school of business that is interested in establishing connections with the small businesses in its area. In fact, funds have already been earmarked for this effort. On your own you have started drafting plans for a Small Business Cooperative Workshop in which student interns would earn course credit by working as consulting operations analysts in small businesses.
>
> The idea first came to you when you were working on graduate placement last year and discovered that your graduates were having trouble getting jobs in small businesses in the area. In subsequent months your consulting work with Mylo Industries led you to wonder if MBAs seemed overqualified to some companies. Many students, whatever size operation they go into, could use practical experience, and your inquiries indicated that businesses often don't realize how they might benefit from students with graduate management training. Later, to research your hunch, you talked informally with personnel at a local printing company, a leather goods manufacturing company, and a local department store. These talks suggested that the image of your graduate school leads business people to expect its graduates to have a highly theoretical, "academic" orientation.

TO: Dean Whatsthepoint

FROM: A. P. Writer

DATE: March 5, 1980

SUBJECT: Talk with Harlan Miller of Handyman

I am going out of town tomorrow and wanted to tell you about my talk with Harlan Miller. This afternoon I went out to Handyman and talked with Mr. Miller, the Vice-President, about the possibility of setting up a Computation Center on a shared-time, shared-funding basis. He was courteous, but he felt that setting up the Computation Center was inappropriate for his firm because the Center's research-oriented applications would not be consistent with his company's operating needs. To obtain maximum benefit, a technical sophistication beyond their present capabilities would be required. Instead, management at Handyman wants to implement a reexamination of key personnel and key departments. As I talked with Mr. Miller, that notion made a lot of sense in terms of the kind of business with which Handyman is involved.

I would like at this time to mention a project I have been formulating that I call the Small Business Cooperative Workshop. In this program students would earn course credit working as interns in small businesses where they would perform as consulting operations analysts at no cost to the small businesses. The arrangement would help to dispel the image of our graduate school as highly academic and theoretical. During the last few months I have done a lot of consulting work with Mylo Industries and have talked to representatives from other companies--including a printer, a manufacturer, and a store manager. They think our students are overqualified; they don't realize that their businesses could benefit from the training of our students. Our students could also use the experience. We could ensure that the workshop was well balanced, incorporating students who have various managerial skills, so that most any problem could be tackled. At first, we would have to advertise to get clients, but after a few successful engagements clients would become plentiful. Payment for services rendered could be accepted as contributions to the general fund. Perhaps Handyman could be our first client, and some of our students could assist in that reexamination Miller talked about.

Figure 1. Memo to the dean: Response A.

Because of your interest in small businesses, the new Dean has just asked you to get in touch with Handyman Industries, a highly influential small business with which he would like to establish a working relationship. He suggested that you explore their interest in setting up a Computation Center on a shared-time, shared-funding basis. You haven't yet told the Dean about your Small Business Cooperative Workshop, which you think is a far better plan, but you know he is considering alternative proposals.

When you described the Dean's Computation Center to Harlan Miller, Vice-President of Handyman, he was friendly but firm. He told you, "A cooperative connection with your school is indeed an attractive area for consideration. However, at the present time the research-oriented applications of your computer installation are incompatible with our normal needs; moreover, the technical sophistication required to maximize our benefit from the facility is beyond our present capabilities. It is generally felt by management that efforts would be better directed, in the short term at least, in re-examining certain key personnel and key departments rather than entering into a project beyond our needs."

You are leaving town tomorrow morning before you can see the Dean. Consider the situation and the merits of your own plan, and write a memo to the Dean. You will need three copies of your memo.

Cases are in part a test. The student does come to class with a written memo that is his or her proposed solution to the assignment; however, the class period brings into play an inductive procedure through which the student moves beyond the solution he or she has prepared. Even the good writer becomes more self-conscious about the choices he or she made.

To let this happen, I bring to class two versions of the memo written by previous students. The class is asked to evaluate them as a reader would. Response A, reproduced in Figure 1, is organized around the writer and the writer's discovery procedure. In focus, arrangement, and selection of information it tells the story of the writer's experience over the last few months. Not only does it fail as a problem-solving analysis, but it is ineffective as persuasion because it neglects to provide a set of mutual goals that would link the writer's plan and the reader's needs. Response B, shown in Figure 2, is more successful in these areas, although neither memo is presented as "correct" or "incorrect"; instead, the memos represent alternative solutions to a rhetorical problem.

After the class has formulated standards of comparison and evaluation, which I record on the board, I develop the set of

TO: Dean Whatsthepoint

FROM: A. P. Writer

DATE: March 5, 1980

SUBJECT: Working Relationships with Handyman and Other Local Firms

---

In response to our recent talk about improving working relations with local small businesses, I have discussed the proposed joint Computation Center with Harlan Miller, Vice-President at Handyman. He was receptive to the idea in principle, but he felt that because of the technical sophistication and research orientation of the project, this particular relationship doesn't really fit their needs.

Clearly, however, he is interested in a cooperative connection, and I think we might consider alternative proposals. One possibility that I have been looking into is to set up a Small Business Cooperative Workshop in which our students would earn academic credit while supplying small firms with their services as consulting operations analysts.

Perhaps the chief advantage of this project is that it might solve several other problems, in addition to establishing a working relationship with local firms. It might improve our image in the small business community, where we appear to be viewed as highly theoretical and not particularly relevant to the needs of small business. If our students could apply their training to the specific problems of small businesses, their work would demonstrate some clear, practical benefits of management techniques. At the same time, the arrangement would give our students practical experience in the field, experience that would not only help them make career decisions but improve their chances on the job market.

Initially we might try to work with Handyman on its reorganization problem. That step would establish a working relationship and might lead to future connections, including the Computation Center. Perhaps we could talk about this when I get back on Wednesday.

Figure 2. Memo to the dean: Response B.

criteria given below. Students then exchange their memos for group grading. First, three students, working individually, use the twenty-point scale to evaluate the same memo (each student has brought three copies of his or her memo to class). Then, working as a committee, the three discuss their decisions and negotiate until they arrive at a single score. The individual grades written down before the committee discussion force a serious re-evaluation by all three graders.

> *Subject heading.* Would it clearly identify the issue in six months when the present context has been forgotten? (2 points)
>
> *Initial paragraph.* Does the writer set up a problem, focus on issues, and get to the point the reader is looking for? Are a problem and a purpose evident, or is the beginning a narrative focused on recent and impending events? (5 points)
>
> *Transition.* How does the writer handle the transition from the Dean's plan to his or her own? Are the two plans connected or polarized? Is a mutual goal established? (5 points)
>
> *Presentation of the plan.* Is the presentation itself focused on solving a problem or meeting a mutual set of goals? Or, is it a discussion structured, like a textbook, around the plan itself or, like a narrative, around the writer's process of inquiry? (5 points)
>
> *Conclusion.* Does the writer open the door to further action? Does the writer plan ahead or simply define and drop his or her "good idea"? (3 points)

One of the chief strengths of a rhetorical case is that, unlike many general cases, it allows direct and explicit evaluation. Not only can the student directly compare his or her solution to those of others in the class on a closely limited set of information, but the class can isolate and discuss specific writing techniques and the features of a good solution.

The rhetorical case has one final value to instructors of professional writing. It asks us to define the skills we think writers in the professions must have. It forces us to identify those intuitive criteria we use as experienced readers. For the student, the rhetorical case provides a laboratory for learning and testing those skills in an authentic and complex, yet teachable, situation.

Notes

1. The question is addressed by Earl W. Britton in his article, "What Is Technical Writing?" *College Composition and Communication* 16 (May 1965): 1-4; and by John Harris in his paper "On Expanding the Definition of Technical Writing" delivered to the Teaching of Language and Literature Section of the Modern Language Association Convention, December 1977.

2. Sharon Crowley also alludes to this method in her article "Components of the Composing Process," *College Composition and Communication* 28 (May 1977): 166-69.

3. For further explanation of this approach, see Lloyd Bitzer, "The Rhetorical Situation," *Philosophy and Rhetoric* 1 (January 1968): 1-14.

4. Wayne Booth's discussion of rhetorical stance is found in *College Composition and Communication* 13 (October 1963): 139-45.

5. Uses of internships at the University of Michigan are described by Ben Barton and Marthalee Barton in "Toward Teaching a New Engineering Professionalism: A Joint Instructional Effort in Technical Design and Communication," and by James P. Zappen in "A Mini-Internship in a Professional Writing Course." Both articles appear in Thomas M. Sawyer's *Technical and Professional Communication: Teaching in the Two-Year College, Four-Year College, Professional School* (Ann Arbor, Mich.: Professional Communication Press, 1977).

6. For a discussion of simulation in the technical writing classroom, see Peter Klaver's article, "Writing as Engineers and Writing in Class: Simulation as Solution and Problem," in Sawyer's *Technical and Professional Communication*.

7. Richard Young, Alton Becker, and Kenneth Pike propose this use of the psychology of Carl Rogers in their text *Rhetoric: Discovery and Change* (New York: Harcourt Brace & World, 1970).

4 Simulation and In-Class Writing: A Student-Centered Approach

Colleen Aycock
University of Southern California

Weekly in-class writing provides students and instructors with on-going evaluations of student writing skills; in addition, it prepares students to write effective out-of-class reports. To provide realistic writing situations and subject matter for these assignments, in-class simulations can be staged. This paper describes one such simulation and discusses the merits of in-class writing based on simulations.

Most teachers of technical writing assign several out-of-class reports that require students to spend many hours defining and researching topics and even more hours writing and polishing papers. Although out-of-class reports are valuable, teachers and students also benefit from more frequent checks on student progress through the writing of weekly reports in class.

The first question facing the instructor who assigns these reports is, "About what shall students write?" After arduous attempts to locate subject matter that would interest my students and provide examples for the specifics of my class lectures, I decided to ask students to share the task. I made the following announcement at the first class period: "A portion of the class time in this course will be devoted to in-class simulations and report writing. Each of you will work in a small group with students who share your professional interests, and each group will illustrate through role playing a situation or issue that might be encountered in that field. During this simulated presentation, observers will carefully document what they see and hear so that they can draft a report in class."

Simulation and report is a versatile scheme. In heterogeneous classes, students can be grouped according to their majors; in more homogeneous classes, they can be grouped according to more specific interests or specialties. Each week brings a new report, and if three class periods per week are available, the first can be given to a skill lecture, the second to the simulation and in-class report, and the third to a discussion of the report. If two class periods comprise the week, the first can be used to discuss the previous week's report and to introduce new techniques or forms; the second can be given over to the simulation and report. If students meet only once a week, the order within the three-hour block of time remains essentially the same.

But where, we might ask—because students inevitably will—is the content for such simulations to be found? Frequently, of course, the instructor offers suggestions, but students who have work experience often suggest ideas to those who have had no experience. Invariably, some students go into the field to ask questions, but this too is valuable. And some decide to discuss current community issues. Recently, for example, a group of geology students presented simulated testimony by assuming the roles of geologists attending a hypothetical hearing on a real proposal for a new dam site on the Colorado River.

A five-minute limit on simulations forces students to choose information carefully. And the writing time might initially be limited to thirty minutes and then reduced as students become more accustomed to writing in class to as few as ten minutes. Time limitations are useful, I think, because they parallel those of on-the-spot communications that most professionals encounter. "How many hours at home," you might ask students, "do you think professionals devote to writing plans, drafts, and revisions of communications they are responsible for during the work day?" Rookies may be slow, but experienced professionals know that economy is measured in minutes, and they learn to write quickly and efficiently during regular business hours. Pressures under time, we all know, exist, but pressures reduce when skills improve.

Oral simulation and written response not only mirror real work experiences but are one of the easier methods for bringing material from the students' major fields into the technical writing classroom. Then, too, the weekly simulation report allows students to experiment with a variety of content and report forms so that they are prepared to write out-of-class reports of greater breadth. In the laboratory setting of the classroom, the instructor illustrates the writing process and asks students to apply these techniques, as opposed to merely lecturing about writing skills and then asking students to solo.

Let me illustrate this procedure with a simulation based upon a company's proposal for flexible scheduling. One member of the student group had given an oral briefing, complete with diagrams, that explained the concept of flexitime. Then, students received a copy of an initial proposal that the fictitious Hal Luben, ad hoc committee chairperson, had sent to the equally fictitious company president, Mr. J. T. Westmont, requesting a trial period for the flexible work schedule.

The simulation dialogue reproduced below opens with a committee seated around a conference table, awaiting the arrival of Brian Wagner, the production manager. Each member has a name plate for ease in identification: Hal Luben, Ad Hoc Committee Chairperson; Stan Glazebrook, Research and Development; Brian Wagner, Production; DuRoss O'Bryan, Comptroller; Gretchen Hallquist, Personnel; Steve DeAngelus, Data Processing; Mike Kolbeck, Supply.

> Brian: Sorry I'm late. We were having quite a bit of trouble making some of the adjustments on that new number 3000 machine of ours.
>
> Hal: Have a seat, Brian, and let's get started. As you know, the purpose of this final meeting is to document each department's position on flexitime and to submit our conclusions in the form of a recommendation to Mr. Westmont for final approval. Brian, I know how busy your schedule is, so why don't you give us your feelings on the matter in case you are called back to the line?
>
> Brian: OK, Hal. Basically our department just won't be able to use flexitime. We're a production department that depends on each one of us being exactly where we're supposed to be, when we're supposed to be there. We can't be wondering when someone is going to show up. Unless . . .
>
> Hal: Unless what?
>
> Brian: Unless management wants to give me a half-dozen or so more people so that I will be able to participate in such a fine program. [The Committee laughs.]
>
> Hal: What about you, Steve?
>
> Steve: We're all for it! We've been waiting for something like this for years. I feel that it will not only improve morale but raise productivity as well. The only place we won't be able to use flexitime is in our Operations Branch. The keypunch and computer operators there are production-oriented and as dependent on each other as people in Brian's department.
>
> Hal: Then basically you're for flexitime?
>
> Steve: That's right.
>
> DuRoss: I just don't see how we can go on a loose time schedule such as flexitime without some other kind of time control. If this

program is approved, I am going to have to insist upon time clocks for everyone.

Stan: Come on, DuRoss, we're not all kids here! My people resent the use of time clocks. They say it makes them feel as if they are constantly being watched. Besides, our department has been successfully using the sign-in method for the last six months. You haven't said anything about time clocks before.

Hal: Let me interrupt. What we want to do here is to establish whether or not we are for the program. If it's accepted, we can discuss necessary controls at a later time. What is your initial stand here, DuRoss?

DuRoss: We are for it, assuming the needed controls are provided.

Hal: OK, and you, Gretchen?

Gretchen: As you know, it was our department that forced the issue with union support when it looked as if management was going to drag its feet. Everyone is for it. It improves morale and production. Just look at the benefits we found during the trial period: less tardiness and fewer uses of sick leave. Some even feel that it reduces travel and parking congestion. And with the school year having just begun, some of the mothers in my department like being home earlier in the afternoons.

Hal: Thanks, Gretchen. Mike, I think you're the only one we haven't heard from yet. What are they saying over in Supply?

Mike: We like the idea very much. But we are somewhat split over its application.

Hal: How's that?

Mike: Well, we feel we can use it in our administrative areas, but we're going to have to exclude Shipping and Receiving. They are production-oriented and have to be there ready to unload when the trucks come in. That's where the problem is. I can see the tension building already between Shipping and our other branches. How are we going to handle something like this, anyway?

Hal: That will be difficult, but I believe we'll be able to solve these problems as have other companies using flexitime. To recap: it seems to me we have everyone, at least to some degree, in favor of flexitime except Brian in Production, Mike in Shipping and Receiving, and Steve in his Operations Branch, where they feel that the concept can't apply. Is that right? [All agree.] Well, then let's adjourn and get back to work. Thanks again for your help.

With this background, students were asked to draft a follow-up memo from Hal Luben to President Westmont detailing the committee's response to flexitime and to formulate, based on that response, a recommendation on the use of flexitime within the company. Thus students would be observing, defining, classifying, generalizing, and verifying.

Although the purpose of the memo had been specified by Hal in the simulation, it was the responsibility of each student to formulate the recommendation and to include appropriate supporting detail. Figure 1 shows what one student wrote in approximately twenty minutes.

During the discussions that follow each written report (and you can see by the example that there is plenty to discuss), the instructor is able to evaluate not only the students but also his or her material and teaching effectiveness. For these reasons, simulation and in-class writing have implications for all writing classes on all levels.

In the same way that this strategy reverses the student role by sending students to the front of the classroom, it alters the instructor's role by bringing him or her back into the class. And it personalizes instruction in other ways. Unlike the individual and lonely frustrations that students encounter when writing out of class, in-class writing gives them immediate instructor response at the time they are facing particular writing problems. As consultant or tutor, the instructor can reopen writing routes or break the block of frustration. In this way, simulation and in-class writing help to equalize classroom roles or at least to make them proportionate.

Of course, this ever-changing laboratory in the classroom makes demands on the instructor, but it also allows for individual differences among instructors and their curricular priorities. The series of simulations must be backed up by lectures that help students handle different kinds of material in different technical forms. For example, documenting the consensus of department heads on the use of flexitime required students to categorize and compare details in an arrangement suitable for an internal memorandum. Indeed, each writing assignment readily lends itself to the introduction of one or two skills that best meet the student's needs for a particular simulation. Depending on the assignment, the instructor might, for example, introduce such techniques as *technical abbreviations* for ease in writing, *appositives* for ease in defining, *free modifiers* for syntactic variation, *parallel constructions* for handling material of equal weight, *transitional statements* for smoothness, *intentional repetition* for emphasis, or *verbal style* to replace the lifelessness of a nominal one. And these are skills that can be stated in performance terms: "After the lesson on ＿＿＿＿＿, the student should recognize that ＿＿(skill)＿＿ best carries the information presented in the simula-

From: Hal Luben, Chair, Ad Hoc Steering Committee

To: Mr. J. T. Westmont, President, Mardex Manufacturers,
           Los Angeles, CA

Via: Mr. David Jeffreis, Plant Manager, Assembly Operations,
           Flagstaff, AZ

Date: September 20, 1980

Subject: Disposition of Flexitime

As a follow-up to my memo of September 4, I can now report that
flexitime will benefit only certain departments within the company
and that it should therefore be adopted only by some departments.
It should not be adopted by all departments. Specifically,
flexitime should not be adopted in Production or in Shipping.

The September 13th meeting of the Ad Hoc Committee found this
flexible working schedule advantageous for administrative personnel
but inoperable for the remainder of the plant employees.

Specifically, those departments of Research and Development, Data
Processing, and Personnel favored the schedule in light of its more
desirable working conditions because it seemed to improve morale
and attendance.

Our Comptroller and Head of Production, on the other hand, found the
proposed schedule costly and ill-suited to production demands. In
favor of the new system, yet feeling that it allowed workers to
take advantage of the circumstances, Mr. O'Bryan was concerned
about the detection of false entries on time sheets. Mr. Wagner
felt, too, that the production unit would not function as well on
such a schedule.

Shipping head, Mike Kolbeck, thought that flexitime would work for
his clerical staff, but was, for his shippers, impractical since
orders and deliveries require a rigid schedule.

To avoid the internal difficulties that would arise if the entire
company were on a flexible schedule, and at the same time, to
take advantage of the benefits it offers, I recommend that the
Departments of Research and Development, Data Processing, and
Personnel, all independent of other department working schedules,
be allowed to adopt flexitime.

Figure 1. Student memo in response to flexitime simulation.

tion and be able to use that skill in writing the assigned report."
As the semester unfolds and the series of simulations continues,
students progress to a more skillful handling of material. And
with each new lesson they should be held accountable for the
previous ones.

In many ways, students are less likely to find the course boring:
the subject matter changes from class to class, as do the roles of
instructor and student. What is offered is a series of experiences,
opportunities for students to observe facts and interpret them in
a meaningful way for specific purposes. Whether writing a technical
description or a procedure, students learn the value of efficiency,
precision, and economy—to avoid the ambiguous statement and
to write as well as they can in the shortest amount of time.

If we have students practice these skills in class before they
work on out-of-class reports, we might be surprised and pleased
by the quality of their independent writing later on. In class or
out of class, they will find that, regardless of the length of the
report or the amount of time they have to write it, the same
basic questions must be asked: What is the focus of this report?
What information must be included? What information can be
excluded? What information is of lesser importance and can be
subordinated? What is the most effective statement for these
ideas? And, what is the most effective pattern of organization
for the facts as I interpret them? In-class simulations followed
by in-class writing, in short, reinforce the instruction of conventional out-of-class writing assignments and bring the technical
writing course closer to the writing situations that professionals
face.

5 Group Projects in the Technical Writing Course

Gerard J. Gross
Locus, Inc., State College, Pennsylvania

Training students to write technical reports as a group project requires them to handle problems they will encounter in future job-related situations: organization of a complex subject, critical evaluation by peers, and deadlines. A class is divided into two research and writing teams that collaborate on projects derived from career interests or local problems.

During my first several years teaching technical writing, I was often troubled by a large class of writing problems and applications that was not touched on in the course as I was teaching it. None of the texts I knew treated this aspect of writing in any detail; yet it was one that in my experience would likely take up a large proportion of my students' time when they faced writing assignments in their technical careers. The aspect of writing I am referring to is the composing of a fairly lengthy technical report as a group project, in cooperation with other workers and writers. I will explain in more detail later why I believe a facility in this type of writing is important, but for the moment I want to emphasize that such work is frequent in many technical careers and that it poses unique problems of style and organization. In order to help students meet these problems, I planned a technical writing course that included group writing projects. I should add that group projects do not comprise the entire course; many of the standard technical writing topics are covered and a number of single-author assignments are given. Yet, this approach does not consist solely of a couple of assignments to be plugged into any course. Rather, the group approach applies not only to the writing of certain assignments but also to their evaluation and to the way a large portion of the course is run.

Though my initial motivation for group projects was highly practical—to give students experience in a kind of writing many would need later on—I discovered benefits that extend well beyond this initial purpose. The dividends in terms of class interest and participation are high, and the skills students learn often apply equally well to areas of writing that are not specific to group projects. I am convinced, therefore, that what I am doing is not only practical but pedagogically sound, perhaps in part because it *is* practical.

The use of group projects as discussed here might be seen as a sophisticated application of a teaching technique called collaborative writing, a technique that has been developed and written about extensively. I do not intend to cover here the general theory and application of collaborative-writing techniques but to emphasize the application to reports with more than one author. For further information on collaborative writing, I refer the reader to Bruffee, Ellman, Gorman, Hawkins, Hoover, Moffett, and Snipes in the bibliography at the end of this collection.

Rationale

I can speak from experience that group writing projects are common at the professional level in industry and government, since I worked six years for Aerospace Corporation, a "think tank" supporting the Air Force's space and missile efforts. I not only wrote a great deal, but I also saw an enormous number of reports of various kinds, written from both the government and the private industry sides. One of my clearest recollections is the large number of reports written jointly. Requests for proposals, proposals, test plans, test reports, and contract final reports are all more often than not written in collaboration with other workers. Even independent research is often done by two or more people, and the report on it is a group effort. While I was at Aerospace, four out of my five major published reports were written with several other people. This is not to mention numerous contributions to work statements, minutes to meetings, and department progress reports, all of which had multiple authors. Certainly much informal intracompany writing is done independently; but in the fields of work with which I have experience, a large proportion, perhaps as much as fifty percent, is collaborative.

Though my experience was in government-related industry, discussions with people in other fields indicate that this experience is representative.

What, then, are the difficulties with group writing that do not arise in single-author work? The problems can be broken down into three categories: organization, style, and scheduling. Perhaps most difficult and distressing is the fact that the requirements for a good joint report place competing, seemingly mutually exclusive demands on the writers. A group report requires careful coordination and cooperation to ensure that it is well organized and that one part is consistent with another in scope and style. Yet such reports must frequently be written within tight schedules, schedules that almost preclude spending the amount of time necessary for a smoothly unified final report.

The process of writing a large proposal illustrates these difficulties well. Some proposals for major contracts are monstrous affairs in which fifty to a hundred people have had a direct hand. Tasks are often closely interrelated, and if one person promises to perform structural analyses on five design configurations, while another proposes to supply material property data for only four configurations, the proposal is in trouble. Each task is often written by the person who will perform the work. If one writes a section in an expansive prose style while another presents a section in outline form, the work proposed may be consistent, but the proposal will read like the patchwork that it is. Deadlines are often extremely rigid, and the time allowed not sufficient to do as extensive a job as one would like. Yet, if the deadline is missed, the proposal is simply not considered, and much company time and effort have been wasted. Some proposals I have reviewed showed obvious signs of haste and lack of coordination with handwritten pages and hastily sketched graphs photocopied and stapled in. Now, work on group projects in a technical writing course may not make every student a whiz at meeting these difficulties, but such work is time well spent. Skills developed in the classroom will be appreciated later when writers are faced with the stress of a very real and demanding schedule.

Application

In using group reports for a technical writing class, my idea is to have students work on topics that are, first, *real*; second, extensive enough so that a group of eight to fourteen students can work on each project; and third, in the ideal case, complex enough so that students in the group can apply some of what they've learned in their majors to the project, yet not so complex that they cannot handle the project in the time available. These projects work best with juniors and seniors who have had some upper division courses in their majors.

I have been splitting my classes (maximum: twenty-eight students) into two approximately equal groups, so that we have two separate projects going in each class. There might be advantages to smaller groups, but it would be difficult for the instructor to keep up with a large number of projects.

So far I have used the technique in two ways. The simpler is to have each group work on a proposal for a large-scale research project or feasibility study. We don't, however, actually carry out the proposed study. This method has the advantage that the project can be ambitious and provide scope for a number of disciplines. It is also relatively easy to complete such a proposal during the course. The disadvantages include the difficulty of keeping the work proposed at a realistic level when students know they won't have to do the work they've proposed. Students who have little or no work experience are particularly idealistic about what they think can be done with a certain amount of time or money. This disadvantage can be overcome to some extent by coordinating the project closely with one or more faculty members from key technical disciplines. Another disadvantage of assigning only a proposal is that students who put so much effort into writing a proposal may feel let down when they see no fruits of their work. Nevertheless, assigning a proposal that will not be carried out works well for instructors who want to include a limited amount of group writing in a course or who want to familiarize themselves gradually with the technique.

The second method I've used is to have students early in the course write a proposal for a more modest research project and then carry out the work and write the final report on it. One advantage of writing both a proposal and a final report is the satisfaction students gain by accomplishing the goals of the proposal. Another is that the work verifies to some extent the merits of the original proposal. Students also get more practice in collaborative writing by doing two reports instead of one. An obvious disadvantage of carrying out the proposal is the amount of work to be accomplished during a three-credit course. Another is that this approach carries students well beyond the normal

boundaries of a writing course. Some students may even resent
having to do the research as well as write about it. I do not recall,
however, having heard strong objections on this score; in fact, I
have received better student evaluations when the course was run
this way. A situation where a homogeneous group has been
assigned a project in another technical course might be ideal;
students would earn credit for their technical work in one course
and for their writing in another. Such a group would also have a
clearly identified monitor for the technical aspects of the project.

Let me, now, discuss how the group writing part of a course
might be run. I will describe mainly the procedures for group work
on proposal writing, but these techniques apply, *mutatis mutandis*,
to the final project report as well.

The first job is to select topics to work on. I usually ask each
student to suggest a topic in writing, with comments on how it
might be developed and on what skills will be needed. Then, after
I have organized the suggestions, grouping together similar topics,
the class chooses two topics to work on. I retain the right to veto
a topic that is popular but impractical. One difficulty is arriving
at topics that will use the talents of as many people in the class
as possible; therefore, I provide students, before they suggest
topics, with a list of the majors of everyone in the class. Here is an
example of the makeup of one recent class, which I divided into
departments as if it were an actual company:

> A Department of Environmental Resource Management, with
> six people in environmental engineering or resource and
> wildlife management
>
> A Life Sciences Department, with six people in animal science
> and plant science
>
> A Forestry Department, with four people in forest science
> and forestry
>
> An Engineering Department, with four people in mechanical,
> industrial, and architectural engineering
>
> An Earth Sciences Department, with two people in geology
> and petroleum/natural gas
>
> A one-man Math and Computation Department

I might mention that the problem of finding a role for everyone
can be alleviated by assigning some people to special duties such as
working with the introduction, conclusion, or flow charts. Incidentally, I generally ask one senior member of the class to volunteer as the coordinator for each project. One duty of this person

is to help run discussion groups after the class has been divided into two groups and I can be with only one.

To give an idea of typical topics—what I mean by *real* projects— here is a list of projects for which my students have written proposals but not carried out the work proposed:

Solar energy applications to housing in the State College area

A feasibility study for an aquarium at Penn State

A feasibility study for a large "Technical Service" building for technical students in all disciplines at Penn State

A plan for reintroducing wildlife species that have disappeared from Pennsylvania

Projects that students carried through to completion and wrote final reports on include these:

A feasibility study of solid waste management in the State College area

A plan for the reclamation of a strip mine area north of Altoona, Pennsylvania

Effects of large clearcuts on the forest ecosystem of Pennsylvania

Specifications for central Pennsylvania housing that will use minimum energy resources

Effects of population density on Spring Creek and ways to improve its water quality

A feasibility study of wind power for generating electricity in the State College area

Note that all topics bear relation to the school, city, or state area. This is not a required attribute, but topics that engage students in problems that touch them in some immediate way are the most successful. Environmental issues are particularly popular and can accommodate a wide range of technical specialties.

After topics have been selected, I tentatively assign students, based on their majors, to one group or the other, allowing them to trade around later if they have other preferences.

We next develop outlines, both for the tasks to be performed and for the report to be written. We usually begin with a brainstorming session in class, listing ideas at random on the board and arriving at tentative outlines by the end of class. This session is often an eye-opener for students who assume that the organization of a research project or report springs forth full-panoply from the mind of its originator. After this class, I go over the tentative outlines with the project coordinators, refine them, and have dittoed versions ready for the following class. During that class, we make additional changes or refinements and are ready to assign students to specific tasks within projects. Because of the nature of the work, outlines, proposals, and final reports are usually modular, or arranged according to discrete tasks.

The process of organizing the work allows students to apply some of the principles of outlining—a topic that of itself can seem remote and uninteresting. This work also confronts students with the problem of outlining a topic of greater complexity than many have faced in individual writing assignments.

When the general outline for each group has been agreed upon and students have been assigned to tasks, they work on a rough draft of the proposal. Ideally, all students in a group would have a chance to look at the entire rough draft of the proposal before the final copy is written, but in practice this has been difficult. Instead, I've asked students to coordinate their work—both in content and style—with those whose tasks bear most closely on their own. Then each hands in a rough draft of sections describing his or her tasks. After I assemble the entire proposal draft, I spend a class period with one group (giving the other group a free period), going over the proposal from beginning to end, covering all major items in each task, and trying to iron out inconsistencies and areas of overlap. This session also gives students a chance to note differences between sections that were well prepared and those that were vaguely stated or poorly written. After we've reviewed the rough draft, students prepare final copies of their sections. This writing is not done in class, but on their own.

The next step is one of the most important in my approach to collaborative writing. When each proposal has been written, it is reviewed by the other group in the class. I try to set up the proposal review process to resemble the way it would be done in industry or government. Students are asked to review the proposal as though they were to decide whether or not the company should get a contract for the work proposed. I assign three or four people —according to their technical backgrounds—to evaluate each section. Students then comment in writing on the content and style of the sections assigned them and on the proposal as a whole. I eventually give these comments to the original authors of the sections. In addition, criteria can be evolved and a point system followed. The five criteria below, for example, are weighted on a 100-point scale.

Understanding of problem (25 points)

Soundness of solution (25 points)

Program management—schedules, coordination of tasks (20 points)

Quality of writing (20 points)

Capabilities of personnel (10 points)

After the proposals have been read and evaluated, we meet in separate groups, this time to discuss the other group's proposal, again from beginning to end. As many teachers have discovered, students are often much better critics than they are writers. In the proposal review, students may be able to identify key problems that they were unable to see in writing their own sections, even after the review of their rough drafts. With proposals, students often find it difficult to distinguish between being specific about what they are *going* to do and the process of actually doing the work and reporting on it. In their writing, they may be totally general about what they will do, or (less often) they may go into such detail that there's no work left to be done in the project itself. A typical reaction from students at the end of a proposal review is, "Did the other group tear apart our proposal as thoroughly as we did theirs?"

So far, this discussion has concentrated on proposal writing; but when students complete a study and write a final report, I follow the same basic procedure. During the research, the main concerns are to see that the work proposed gets done and that there is a timely flow of information. For example, it may be necessary for one person to finish determining how much solid waste is generated in the State College area before the sizing requirements (and hence cost) for a reclamation system can be laid out. Occasionally there are real snags in the way of timely completion of the work proposed—for example, failure to receive information requested by mail. A quarter or even a semester time span can be very limiting when outside information must be sought. In some situations, the instructor and the group must agree upon a modification of the proposed work. Periodic group meetings to discuss progress and to work out difficulties are helpful. Incidentally, projects such as these provide ideal opportunities for students to practice writing progress reports, since now they actually do have subjects, and hopefully progress, to write about.

As the due date for the final report approaches, I begin emphasizing design and layout, accounting for modifications from the original organization of the proposal. The writing procedure follows the scheme I have described for proposals: students prepare rough drafts of their respective sections, meet to coordinate the parts, write a final version, and then have that version reviewed by the other group, with written comments on each section by several students. Since this work comes near the end of the term, it can be difficult to find the time to carry out the review process thoroughly. I find it wise to set up and maintain a rigorous schedule of due dates and review sessions, since the review by the other group is such an important part of the learning process.

A word or two about grading. The degree to which student reviews and comments are factored into the grading of assignments depends on the individual teacher. I keep the review process and my grading basically separate; that is, I assign a grade to each person's contribution that is independent of the comments by students on that section (this is not to say that our opinions will always disagree). This method leads to much freer discussion and to a greater willingness for students to point out real deficiencies in the work of others. I do, however, assign a proportion of the final grade in the course, perhaps twenty percent, to participation (including group meetings during the work on the project and reviews of the other group's reports) in order to emphasize the importance of this aspect of the course. Conceivably one might assign a single grade to the entire proposal or final report, and give that grade to each person who wrote a section of it. This policy might work for a small group, where the work and writing could be more closely coordinated; but for a large group I feel that the possible inequities outweigh the motivational force of a common grade.

Some Concluding Observations

I have mentioned in passing several advantages and disadvantages of this technique of teaching writing. I would like, now, to review some of these and to comment on what students learn about writing from working on group reports.

First, I find that these projects are highly successful as classroom techniques. Students are usually very interested. They tend to be more motivated to work on a real project than on a made-up topic and are pleased to see a final product of significant magnitude. The projects provide variety in the classroom. Periods devoted entirely to lectures can drag, but a period with a brief lecture on some topic in technical writing followed by group meetings works well. Group meetings also give students a chance to develop important communication skills in addition to writing.

As for what students learn specifically about writing, I see the benefits of this approach coming first from the writing itself, and then from the process of evaluating the reports of others.

In writing of this kind, students do think about their audience, a most important aspect of technical writing or any kind of writing. They are not merely writing a term paper to show off facts; they are communicating with someone. The knowledge that the proposals and reports will be evaluated according to a determined scheme helps to direct their emphasis in writing. In addition, the awareness that the reports will be read by other students, and not just by the teacher for a grade, can lead students to try for a better quality in their writing than they might otherwise attempt. In some cases the finished report is sent to a responsible authority; for example, the Altoona strip mine report was sent to the Pennsylvania Environmental Resource Department.

Important also, as I've mentioned, is the experience of coping with the coordination and organization of a complex subject, both in organizing the written report and in working with others. Writing a section of a longer report forces students to be more aware of transitions, to relate what they say to what went before and will come after. I find that writing a part of a longer report also helps students to be specific about the objectives and purposes of their own sections.

Evaluation by peers is common to any application of collaborative writing. In the types of evaluations I use, there are benefits that are specific to the type of report students are writing—for example, whether a proposal adequately demonstrates the ability to provide a solution to a problem—and more general benefits for their writing style. The benefits also work both ways—for those being evaluated, and for those doing the evaluation. The persons doing the evaluation see a number of different approaches. They may be made aware of certain stylistic techniques that they have not tried in their own writing. Or they may see where others have been much more detailed in content than they, or perhaps less. Those being evaluated have the benefit of comments on their writing from several points of view and are not reliant solely on the teacher's adherence to the "party line."

I would like to call attention to some particular problems with group assignments without, I hope, ending on a negative note. As you can imagine, the quality of writing varies considerably from student to student, despite emphasis on unity of style in a group report. This variation extends not just to the quality of writing, but also to the quality of participation in the projects. Some students do not attend meetings, and their sections are not properly coordinated. Or their sections of the report don't get done on time, and an incomplete report is presented for evaluation. Generally, getting the report coordinated, written, and evaluated requires very tight schedules—which are sometimes not met by everyone. A few people who don't contribute adequately can significantly mar the overall quality of the project. In most cases in my own experience, the reports have fallen somewhat short of their potential.

The problems I've just mentioned, however, are not limited to the classroom. Many of them are *exactly* those met in the real working environment. People *always* vary in their ability to write and in the quality of their participation. True, if workers let you down completely on a job, you can fire them. But there will always be those who are more prompt than others, whose initiative you can depend on more than that of others; and the quality of people's work, even those kept on by a company, varies. Yet group reports—good ones—are important. What I am saying, then, is that even if the problems I've alluded to here are not solved completely in class, the experience with group writing will help equip students to cope with the same problems when they meet them in their professional careers. The very reason that might lead to a decision against working with group reports in a class—the practical problems to be encountered—are precisely the reasons why group reports *should* be tried.

I firmly believe that writing projects such as these are effective pedagogically and are helpful in training students for specific tasks. Students are motivated in class, they develop their abilities through both the writing and evaluating processes, and they prepare themselves for the time in their careers when they will contribute to writing group reports.

6 The Functional Writing Model in Technical Writing Courses

Anita Brostoff
Carnegie-Mellon University

Because of limited time, money, and personnel, technical writing courses are not always available to all students who need them. As a result, writing teachers are sometimes asked to collaborate with technical faculty, to consult with students, and to provide writing assignments or workshops for students enrolled in technical courses. One approach to teaching technical writing in such a collaborative context is described here. The approach emphasizes the organizing idea, the reader's frame of reference, forecasting, and continuous forecasting.

At Carnegie-Mellon University, the technical and professional schools are meeting the growing demand for better student writing by teaching writing within the context of the professional schools themselves. The staff of the Communication Skills Center has therefore become increasingly involved in teaching writing by collaborating with faculty who teach courses in the technical schools. Our problem has been to develop methods and materials appropriate to that context. A solution we have used effectively is based on a functional writing model originated by A. D. Van Nostrand and colleagues.[1] Although the method is sequential and cumulative, we find that juniors and seniors and graduate students often do not need instruction at each step. Instead, we choose particular concepts to emphasize on the basis of individual problems identified through writing samples and on the basis of the general demands of a given discipline. Then we design exercises that require the use of these functional writing strategies and that contain specific content from a given discipline.

Over the past three years, I have applied the essential concepts and strategies of functional writing with students in architecture, psychology, urban and public affairs, history, prelaw, business, and engineering. On the whole, I have found the model to be highly adaptable for the various kinds of thinking and writing that are done in carrying out the world's work. The model seems to work in such a variety of areas because the same core of rhetorical and thinking competencies underlies so many writing tasks: analyzing, making inferences, supporting generalizations, and organizing information and ideas. Additionally, similar fundamental demands exist for most technical and professional writing: thorough consideration of the writer's purpose and the reader's response—how the reader will use the writing; absolute clarity; self-evident logic and organization; sufficient evidence to support generalizations; and concise, precise, mechanically correct prose.

In answering at least a substantial portion of these needs, the functional writing model rests solidly on principles derived from contemporary research and scholarship in learning theory and language behavior. The model is based on the assumption that writing is a process and that writers need to be aware of the steps in this process in order to control it. It looks at writing as a process of making and stating relationships, a process one goes through, ideally, by making reasoned choices about what relationship to state, or to develop, next. Writing, according to the model, is a means of learning about one's subject. And, finally, the model is based on the assumption that the product must be viewed in terms of the reader's need to understand and to use that product. Accommodating the reader's needs, according to the model, helps the writer compose more easily, quickly, and effectively.

Specific aspects of the functional writing model that I regularly teach include (1) the organizing idea—the logically unifying component of a paragraph or a sequence of paragraphs, (2) the reader's frame of reference—defined as his or her perspective or world view, (3) forecasting—predicting for the reader the organizing idea and the sequence in which it will be developed, and (4) continuous forecasting—revealing the direction or line of reasoning in an argument.

The Organizing Idea

The first strategy, the organizing idea, is fundamental, and in a sense obvious (good writers surely have always done it); yet, it is perhaps the most frequently violated element in technical writing. An organizing idea is an assertion that relates and incorporates the information in any paragraph or sequence of paragraphs. Writers create an organizing idea by making an inference, by generalizing about a set of information. Perhaps this principle is so often violated in technical writing because writers tend to assume that their readers know without being told the inferences writers are making. But without an organizing idea the assertions in a written statement remain logically unrelated and the particular significance a writer attached to the information remains a mystery to the reader; as a result, the reader has difficulty putting facts into a context.

Following the functional writing model, however, writers begin the process by listing a set of information they have obtained, usually from on-the-job investigation, and by asserting an organizing idea—what they want to say about that information. Consider, for example, the organizing ideas developed by three engineers from the information on electric cars shown below.

Subject: Electric Cars

Set of Information: limited performance capability, battery power, twenty years to develop, pollution, range of forty miles

Organizing Ideas:
1. The energy shortage that will surely be with us for some time to come may be somewhat alleviated by the reintroduction of the electric car.
2. More spending is expected for electric car development.
3. Despite the efforts of the engineering and scientific communities for twenty years, it is doubtful that an effective electric car can be built to equal or surpass the gasoline-driven automobile.

One can see from these three organizing ideas that no single organizing idea is inherent in the information; therefore, when a writer presents data without an organizing idea, the reader must guess the significance of the data or the context the writer has in mind. This is a concept that students who will write technical communications must master.

The Reader's Frame of Reference

Frame of reference refers to audience, the most important component in the professional communication situation. Writers must analyze their work for the purpose of making a given communication clear to a given audience and for the purpose of accomplishing a given intention—whether that intention is for the reader to understand something, to adopt a particular opinion about something, or to act in a certain way. (The question of the writer's purpose is not stressed in the model as much as I think necessary for technical and professional writing, but it's there and must be pointed out.)

After introducing the notion that different readers have different frames of reference (and ones that differ from those of the writer), the functional writing model emphasizes that writers must learn to describe their audiences according to objective characteristics such as level of knowledge and expertise and—far more difficult—according to subjective characteristics such as political leanings within an organization, personal goals and values, and likely attitudes and assumptions. The example given below, written by a middle management person in a major industrial firm, contains a fairly complete description of the readers that writer must satisfy.

> Readers: Management and staff in corporate planning and comptroller groups
>
> Characteristics:
> 1. Audience level varies from paraprofessional to advanced professional.
> 2. Professional levels in the corporate planning and comptroller groups are not equivalent.
> 3. Reader characteristics vary considerably, based primarily upon how each reader can use the information.
> 4. The writer's discipline is management science, but reader knowledge of this discipline varies from ignorance to familiarity. A need exists to bridge this gap. Ignorance threatens some readers, i.e., some readers view the mathematical model as a "black box."
> 5. There is a need to establish the value of the math model to the readers. The document must present its meaning and value, i.e., how it will help readers do a better job.

Next, the model suggests that writers accommodate the reader's frame of reference by establishing a common ground. The example below, written by a graduate student in urban and public affairs, shows how the same organizing idea was presented to two different readers.

> Organizing Idea: I am seeking support for a project that will investigate the causes of the financial plight of minority-owned businesses and will recommend ways to stimulate and expand these businesses.

Reader A: A minority owner of a thriving business in a minority community

Statement to Reader A: The more thriving stores there are in your community, the more people will shop there. It is good business practice, therefore, to invest time and money in projects to strengthen struggling but potentially successful businesses in your community.

Reader B: A city counselor

Statement to Reader B: Although the initial outlay of tax dollars to struggling minority-owned businesses may seem risky, stimulating and expanding these businesses could create jobs, decrease welfare payments, and increase tax revenues. Therefore, it is beneficial to the city to support projects that help minority-owned businesses.

One sees how each reader can relate to the tailor-made statement and how neither need feel threatened by the proposed action. We often ask students to practice using reader frames of reference by writing short statements of definition or explanation for readers at two levels, as in these definitions of *autoxidation.*

Definition of *autoxidation* addressed to a technician who is a high school graduate but has little previous experience in a chemical laboratory:

Oxidation means the combination of oxygen (or air) with a substance. For example, when coal is burnt in the air to carbon dioxide, the process is called the oxidation of coal. Autoxidation is the term used to describe an oxidation in a chain process. For example, substance A is oxidized to B, which in turn is oxidized to C and D.

Definition of *autoxidation* addressed to a research chemist:

Autoxidation is a degenerative oxidation process. It usually proceeds by a radical process. Radicals combine with oxygen molecules to form numerous radicals. Recombination of radicals results in process termination.

Forecasting

Readers are busy people who want to know quickly, at the very beginning, what a piece of writing is about, how the aspects of the subject will be developed, and what conclusion the writer has arrived at. Indeed, the forecast may be all the reader reads; he or she may need to know only the generalities or conclusions and may skip the particular details altogether. Further, forecasting allows an audience to settle more or less comfortably into a communication with a sense of knowing where it's going; fore-casting enables an audience to put all that follows into a context.

Specifically, the functional writing model uses the forecast as an introduction, requiring students to build this introduction by going through the prewriting activities of grouping information into categories (which will become paragraphs) and sequencing these categories in a logical order. The example below, containing information from an architectural study, shows how forecasting functions as an organizational activity. The categories developed from the information set are numbered to show the sequence in which they will be described in the forecast paragraph and in their subsequent development as individual paragraphs.

Subject: The American Porch

Set of Information:

Most common use of the porch in English colonies during the seventeenth century was as an internal vestibule space.

Dutch builders in New Jersey and New York considered the porch an outside living space.

The large central hall in southern plantation houses originated as an outdoor living space.

The climate, surroundings, and style of living in the South created a house type that was unique in the United States, and the porch was an integral element of that type.

An example of the Greek revival is the Ohio T-shaped plan, which has a porch with Greek columns.

In the late nineteenth century, houses with extensive front porches were built all over the United States to maximize the pleasures of outdoor summer life and to set the stage for neigh-borly interaction.

Bungalows, popular at the beginning of the twentieth century, were a response to climatic conditions.

The porch declined for the following reasons: (1) use of tele-phone for gossip, (2) automobile noise, (3) TV for entertainment, and (4) air conditioning.

Possible Organizing Idea: Porches in America were a response to climate, the style of living, and foreign influence.

Possible Categories and Sequence of Information:

3 Porches were used as inside or outside space.

1 Porch designs were a response to climate.

4 Porches were built for relaxation: the view, chatting with neighbors, etc.

5 Building and use of porches has declined recently.

2 There were foreign influences in the design of porches.

The forecast paragraph itself states the organizing idea and the supporting categories in the sequence they will be developed. Such a forecast, the basic stuff of exposition, works well for professional writing in the social sciences and in some technical writing, but the underlying notion is clearly applicable to the more complex process of forecasting in technical reports—the problem-purpose statement. In the following example of a problem-purpose statement that forecasts, the writer has clearly delineated the necessary parts: the problem, the task, and the rhetorical objective of the report.

> The biodisk waste water treating system is currently being employed effectively to treat waste waters containing high levels of contaminants such as raw municipal sewage and raw industrial waste waters. This study was initiated to determine if this type of treating system would be effective in treating low contaminant waste waters such as the waste water entering the GS&T treating facility.
>
> The purpose of this study was to examine the performance of a biodisk treating system by testing the influent and effluent from the system and comparing the results with those of the more conventional waste water system employed at GS&T. The test results show that the biodisk unit was more effective in removing the contaminants than the present system and should be considered as a technique to upgrade the GS&T system and to insure compliance as NPDES limits become more restrictive.

Continuous Forecasting

Understanding the essential meaning of forecasting helps to understand the concept of continuous forecasting. Forecasting is the technique of initially revealing to readers how the writer intends to develop the subject; it gives readers a sense of direction. Continuous forecasting means that the writer continues to indicate to readers the direction of the line of reasoning by showing how each new assertion relates to what has come before. Revealing the pattern of the line of reasoning helps both writer and readers anticipate what will come next and makes clear the logic of the relationships among the assertions.

What continuous forecasting teaches is how to achieve coherence. The functional writing model suggests that an argument is coherent to the degree the logic of the sequence of assertions is self-evident. The reader must be able to discern how each assertion follows from the preceding ones and leads to those

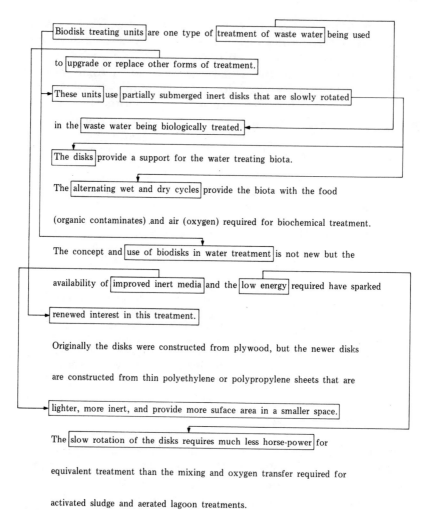

Figure 1. Continuous forecasting: Highlighting the logical sequence of assertions.

following, just as he or she can detect, for example, the logical relationships in the sequence 2, 4, 6, 8, 10. Continuous forecasting, then—making and revealing relationships—assures sequential logic. Successive connections are highlighted in Figure 1 to show forecasting at work.

Strengths of the Approach

In discussing these four components of the functional writing model, I do not mean to imply that they constitute the only significant parts of the model or that they answer all needs in technical writing. Their strength lies in their focus on the reader and on clarity, coherence, and organization. But there are several general advantages in using selected, key concepts, especially when the technical writing instructor serves as a consultant in professional or content courses and is asked to build a writing component into a course or to offer a brief course in writing, even a single workshop. First of all, as consultant instructors we need a set of strategies we can present quickly and fairly simply. Although it makes eminent good sense to teach technical writing in a professional context, the professional content does take precedence and the amount of time that can be devoted to writing is limited. Furthermore, as consultant instructors we need flexible strategies through which we can stress particular areas for particular professional audiences—evidence and hypothesis in social science, for example, or forecasting in the technical report. And we need concepts that give us a specific vocabulary for discussing writing and clear criteria for evaluating writing that can finally help students learn to evaluate their own writing. The functional writing model, in its ability to fill these needs, can become the vehicle of collaboration between teachers of technical skills and teachers of writing. Presenting all or selected parts of the functional writing model, as a situation allows and demands, helps technical writers write faster and better; it gives them strategies they can use.

Notes

1. A. D. Van Nostrand, C. H. Knoblauch, Peter J. McGuire, and Joan Pettigrew, *Functional Writing* (Boston: Houghton Mifflin, 1978).

7 A New Approach to Teaching a Course in Writing for Publication

David L. Carson
Rensselaer Polytechnic Institute

Courses in writing for publication have become valuable additions to increasing numbers of technical writing programs. To make those courses successful, students need broad and realistic experiences that simulate the writing processes of professional writers.

In the past decade technical, professional, and scientific writing programs across the nation have been both improved and enlarged by the addition of courses designed to prepare students to write for and publish in professional, scholarly, and popular journals. Although writing-for-publication courses succeed more often than not, they sometimes incorporate a flaw common to many writing courses. Specifically, they rely too heavily upon a one-on-one relationship between instructor and student during the presubmission phases of writing. Certainly this is not to suggest that close instructor-student association is undesirable; however, that pedagogical strategy may not always be practically sound. Few teachers, however eclectic, are sufficiently expert to judge the publishing possibilities or the technical accuracy of student articles, and few have time to master the stylistic conventions of the many periodicals for which their students may eventually write. As a result, students who write to meet an instructor's standards may ultimately diminish their chances for publication.

To minimize these disadvantages and at the same time to preserve the value of the close instructor-student relationship, I developed a modified approach to teaching writing for publication that emphasizes group involvement as well as self-directed activity and in many ways simulates the writing process of professional writers.

The course requires each student to write three articles and to submit them to journals during the fifteen-week semester. Students choose their own subjects, usually ones with which they are familiar, and submit their articles to journals they themselves have selected. At the advanced undergraduate or graduate level, I require students to direct one article to a scholarly or professional journal, another to a less formal journal of high repute, and the third to any publication selected by the student and approved by me. A chemistry major, for example, might write a first article for publication in the *Journal of the American Chemical Society*, a second for *Popular Science*, and the last for the Sunday science section of a newspaper. A potential teacher of technical writing might try the *Journal of Technical Writing and Communication*, *MORE*, and, perhaps, the Staffroom Interchange of *College Composition and Communication*.

A point I should like to make here is that responsible flexibility is the key. A student might well justify aiming a second article at either *Scientific American* or *Popular Science*, and few would have difficulty in convincing me that a try at publishing short fiction or poetry was not in order.

Few teachers of courses in writing for publication use textbooks because few are available. Although a number of journalism texts may be adapted, few of these are sufficiently directed toward the task at hand. There are, nonetheless, other materials which students find helpful.

I insist that each student own a writing text or English handbook that includes sections on grammar, syntax, punctuation, documentation, proofreaders' marks, and other conventional matters. I also recommend that students own a reputable dictionary and a thesaurus. In addition, I frequently ask them to buy the current *Writer's Yearbook* and often use *Scientific American* as a required text.[1] The former introduces students to the world of professional writing and provides an abridged guide to commercial markets. The latter offers a convenient vehicle for the analysis of style, audience, organizational strategies, and the use of captions and graphic aids.

Beyond this, I frequently ask a reference librarian to conduct a tour of the periodical section, and I place a number of helpful volumes on the reserve shelf, including a recent edition of *The Writer's Market*,[2] several leading technical writing texts, a list of other pertinent reference materials, and a brief bibliography of articles having to do with writing for publication.[3]

WALTER F. DATER UNIVERSITY
School of Humanities and Social Sciences
Department of Technical Communication
Strasburg, Pennsylvania 17579

January 13, 1980

Dr. John C. Pratt
Director Airborne Optics
Watervliet Arsenal
Gap, Pennsylvania 17586

Dear Dr. Pratt:

For a number of years, we have conducted a highly successful course
in writing for publication for advanced undergraduate and graduate
students. The objectives of the course are to introduce students
to the techniques and methods of writing articles on technical,
scientific, and professional subjects of their own selection. The
course's ultimate goal is publication of the students' articles in
reputable journals.

Since students in the course come from a variety of academic backgrounds, we cannot provide adequate expert advice and rely on
volunteer technical advisors to bridge the gap. The responsibilities
of a technical advisor are neither time consuming nor demanding,
and most experts who have participated in our program find the
experience pleasantly satisfying. Should you agree to serve as a
technical advisor, your expertise will be invaluable in providing
guidance on such matters as technical accuracy, timeliness, and
journal selection.

Your participation will assist a student in meeting the real
demands of the professional world.

I thank you for your kind consideration.

Sincerely,

David L. Carson

David L. Carson
Director
Program in Technical
Writing and Communication

Figure 1. Instructor's letter to prospective technical advisor.

Structure of the Course

My writing-for-publication course is essentially "front loaded"; that is, I concentrate lectures, class discussions, and other matters preliminary to the writing of articles during the first three weeks of the fifteen-week semester. Thereafter, almost all class periods are devoted to student-controlled activity.

After the first three weeks, my role becomes that of senior editorial advisor or consultant, and sometimes arbiter. This informal approach within a highly structured system provides advantages not normally attained with traditional pedagogies. It enables me to develop a much greater awareness of both individual and group progress, and it also creates an atmosphere in which my one-on-one discussions with student writers become more effective and, as a result, more frequent. The informality also permits me to interrupt student activity in the classroom whenever the need arises to address problems that are common to the group. When a student's problems are too extensive to resolve in class, I arrange to meet that student in my office.

I employ several techniques designed to give students broad and realistic experience in the process of writing for publication. Among these are guest lecturers (writers, editors, and publishers) and tours of local publishing houses. In addition, I rely on outside technical advisors, oral presentations by students, and student editorial committees. It is to these three that I wish to address the following discussion.

The Technical Advisor

A decade's experience in arranging for students to consult with outside technical advisors during the writing of scientific and technical papers suggested that this method might also work well in a writing-for-publication course. Aside from ensuring technical accuracy, outside advisors provide valuable assessments of the timeliness of a proposed article and its appropriateness for a particular journal. In cases where students are attempting to write popular articles on subjects with which they are not sufficiently expert, technical advisors can be most helpful.

Students normally find their own technical advisors in other academic departments or in the local community, but occasionally they need my assistance. When students confer initially with prospective technical advisors, they take with them a form letter that they have retyped, addressing it to the advisor and securing my signature (Figure 1). This letter explains the nature of the

Speaker _____ Assessor _____

Instructions:

1. Assign the number that best describes the presentation in each category and place that number in the appropriate blank.
2. Write constructive comments for each category in the space provided. If you need additional space, use the back of this sheet.
3. Add the numbers in the right-hand column, recording the sum as Total.
4. Subtract points for distractors, but *only* if you include descriptive comments identifying these distractors.
5. Record the result as Final Score.

Content and Organization: Content was pertinent and organized to permit concise, coherent communication.

Content
21	25	30	_____
fair	good	excellent	

Comments:

Organization
21	25	30	_____
fair	good	excellent	

Comments:

Delivery: The speaker was poised, spoke clearly and with proper diction, showed adequate preparation, and spoke directly to the audience.

17	21	25	_____
fair	good	excellent	

Comments:

Visual Aids: The speaker had prepared well-designed visuals and used them effectively.

11	13	15	_____
fair	good	excellent	

Comments:

Total _____

Distractors   Minus Distractor Points _____

Comments:

Final Score _____

Figure 2. Sample oral presentation critique.

course and the role the advisor is being asked to play. If the technical advisor agrees to participate, the student presents two copies of a letter of agreement addressed to me for the advisor's signature. The student leaves one copy with the advisor and returns the other to me.

Few prospective advisors in or out of academia turn down well-presented requests for assistance, and the results of these liaisons are often beneficial not only to students but to the course, the program, and the department as well. It is not unusual, for example, to find students from the department of a technical advisor subsequently enrolled in the writing-for-publication course or in other courses offered by the technical writing and communication department.

The Oral Presentation

Although oral presentations are not usually considered germane to courses in writing for publication, the oral presentation in this course simulates the proposal to publish that an author might offer to a publisher's editorial staff and much can be said for its inclusion. First, few students have had the opportunity to make a formal presentation on a serious but familiar subject. Neither have they had opportunity to devise visual aids nor, commonly, to express critical judgments about the oral presentations of others.

Although oral presentations may be handled in a number of ways, let me make a few suggestions.

1. Design a critique sheet that enables students to assess the presentations of others quickly and accurately. See, for example, Figure 2.

2. Discuss the dynamics of information transfer through the oral mode, explaining the differences between a presentation designed to convince and persuade on the basis of clear, concise, coherent factual information and one designed to persuade by means of rhetorical devices such as an emotional appeal. I do this by giving an oral presentation *on* oral presentations in which I demonstrate the techniques and methods I expect students to use.

3. Require students to use visual aids such as flip charts to support their presentations.

4. Explain how the critiquing and grading system will function. In my course each student rates all speakers. When the

critiques for a given student are collected, that student's editorial committee, chaired by the editorial advisor, makes a brief oral critique of the presentation. The student's grade is derived by averaging the grades of all student critiques.

Since article deadlines generally fall in the seventh, tenth, and fifteenth weeks, oral presentations take up the better part of classes during the seventh and eighth weeks. Although time limitations generally restrict oral presentations to the first article, the exercise is one of the most valuable of the course, and students generally perceive it to be so.

The Editorial Committee

Increased student involvement begins when I divide the class into editorial committees of up to five students each. These committees do not curtail a student's freedom to discuss work at any stage of its progress with the instructor, but they do formalize a system through which students are afforded a great deal of additional advice not unlike that which they might receive in a professional setting. The typical sequence of activities for writer, editorial advisor, and editorial committee in bringing a first article to the submission point is summarized below.

Writer	*Editorial Advisor*	*Editorial Committee*
Selects tentative subject and surveys several journals for placement potential.	Discusses subject and other preliminary matters with writer.	
Submits prospectus of article with placement analyses of two appropriate journals to editorial advisor and instructor.	Reads prospectus and placement analyses, surveys several issues of journals chosen by writer, discusses matters with writer and committee, and makes recommendations to writer, committee, and instructor.	Reviews subject choice and placement analyses with editorial advisor and makes recommendations to instructor.
Discusses prospectus with instructor.		
With instructor's approval, arranges to discuss prospectus with technical advisor and returns letter of agreement from technical advisor to instructor.		

Submits copies of revised prospectus and letter of inquiry to journal editor to editorial and technical advisors.	Discusses revised prospectus and letter of inquiry with editorial committee.	Makes recommendations to writer through editorial advisor.
	Discusses committee recommendations with writer and sets dates for completion of initial draft and for oral presentation of proposal to class concomitant with submission of second draft (no later than ten days after submission of initial draft).	
Sends revised letter of inquiry to journal editor.		
Begins writing, submitting copies of initial draft as scheduled to instructor, editorial advisor, and technical advisor.	Returns initial draft within twenty-four hours and discusses recommendations for change with writer.	
Writes second draft and prepares oral presentation.		
Submits second draft to instructor and editorial committee.		Reviews second draft and makes recomendations to writer through editorial advisor.
	Reviews second draft and transmits recommendations for change to writer.	
Makes oral presentation.	Chairs critique by editorial committee of oral presentation.	Delivers brief oral critique of oral presentation.
Revises and submits to instructor final draft in envelope addressed to editor of journal chosen for potential placement.	Discusses finished draft with writer and instructor.	

In a sense, each committee represents a journal staff, and each committee member serves as an editorial advisor to a fellow student throughout the process of bringing an article to the submission stage. The advisor also serves as a link between the student writer and the editorial committee and between the editorial committee and the instructor. The list that I prepare to designate editorial committees also assigns an editorial advisor to each student for each article. In practice, then, a student works closely with three editorial advisors and three editorial committees during the semester.

Reflections on the Course

I confess that at times the pseudobureaucratic structure of the course troubles me, but the course has produced results generally superior to those derived from the typical writer-for-publication course. By simulating aspects of the publishing process, it transcends the purely academic context that students tend to associate with writing courses. By incorporating the process of writing into group functions, the course provides students with constructive criticism from several objective sources at several points in the writing progression. As a result, almost from the outset, students look at their own writing much more objectively than they had previously done. This perspective is enhanced by the fact that each student simultaneously serves as an editorial advisor for another student. When students analyze each other's writing from standpoints of correctness, clarity, conciseness, and coherence, they become more conscious of these qualities in their own writing. And when they learn that each criticism of a piece of writing requires a clearly justifiable explanation, they are less apt to construe criticism of their own writing as mere opinion.

All of this tends, finally, to coalesce into a single reason for the system's success: greater, and more active, student involvement in the writing process. During at least eighty percent of the class periods, students are engaged in self-directed activity having to do with some phase of the writing-for-publication process. *Process* is the key word here because writing and publishing, like learning itself, are processes, and efficient mastery of process requires active engagement. I suggest that methods such as the editorial committee are effective because they act catalytically to develop within students an affective knowledge of the writing process in contrast to the merely cognitive perception. The immersion of students in the writing process through action-centered learning situations produces, in my opinion, marked improvement in their attitudes toward their own writing and discernible improvement in the quality of that writing.

Notes

1. *The Writer's Yearbook* is published annually by F&W Publishing Company, 9933 Alliance Road, Cincinnati, Ohio 45242. For another discussion of *Scientific American* in the technical writing classroom, see Wayne A. Losano, "*Scientific American* in the Technical Writing Course" in this collection.

2. *The Writer's Market* is also published by F&W Publishing Company.

3. The best continuing bibliography is *Annual Bibliography of Technical Writing*, edited by Donald H. Cunningham and published each fall since 1975 in *The Technical Writing Teacher.* See also "A Bibliography of Resources for Beginning Teachers of Technical Writing" by Carolyn M. Blackman in Thomas M. Sawyer, ed., *Technical and Professional Communication: Teaching in the Two-Year College, Four-Year College, Professional School* (Ann Arbor, Mich.: Professional Communication Press, 1977). One of the more useful bibliographies for the teacher of technical writing was prepared by Karen A. Edlefsen for my 1978 writing-for-publication course and published in the spring 1979 issue of *The Journal of Technical Writing and Communication.*

Part Two: Components

8 From Aristotle to Einstein: Scientific Literature and the Teaching of Technical Writing

Stephen Gresham
Auburn University

Many students in technical writing classes are absorbed in the immediate developments within their technical disciplines. However, scientific and technical literature of the past can be used as a base upon which to build assignments that serve technical writing class objectives. A series of such assignments is outlined here.

The philosopher Owen Barfield has said, "There may be times when what is most needed is, not so much a new discovery or new idea as a different 'slant'; I mean a comparatively slight readjustment in our *way* of looking at the things and ideas on which attention is already fixed."[1] His comment, it seems to me, applies to those of us who teach technical writing, for we often turn to our dusty files, hoping to find something stimulating for our course only to come away disappointed, if not empty-handed. While offering no dramatically new idea here, I will outline a series of exercises that center on the use of past scientific literature in the technical writing classroom. In short, I offer a different slant.[2]

The skeptical instructor might question whether scientific and technical writers of the past produced exemplary prose. But they did! In fact, in *The Beginnings of Modern Science* Holmes Boynton claims that these earlier writers "for the most part wrote more interestingly about their work than do the highly accomplished specialists of more recent days."[3] He adds, "Galileo, Newton and Franklin were speaking to the world, and let themselves be openly enthusiastic about their discoveries. So they wrote freshly, in a style that for the most part the world can understand."[4]

Yet, in spite of Boynton's enthusiasm for scientific writers of the past, our courses must address the immediate and practical task of helping students write clearly and precisely. I submit that the assignments described below will do that—and they will go several steps beyond; but first the instructor should orient the class to the major objectives of studying past scientific and technical literature.

Objectives

I have identified the following five major objectives for my technical writing students and find it beneficial to spend most of one class period discussing them.

1. To demonstrate to the student that technical writing is a discipline with a history and a solid tradition.

2. To enable the student through the judicious reading of scientific literature to use the following rhetorical modes effectively: classification and analysis, process description, description of a mechanism, definition, and argumentation.

3. To demonstrate to the student that good scientific and technical writing is clear, precise, and audience-oriented. (Earlier scientific literature is especially helpful here because it was generally written for lay audiences.)

4. To introduce, or perhaps reintroduce, the student to the "poetry of science"[5]—the notion that science, as Frederick Houk Law has said, "interprets life and reality as no mere abstract philosophy has ever been able to interpret them. It reveals the world in the deepness of its wonder, and in the fulness of its power. It awakens men to passionate interest in all things, from the grain of dust to the unbelievably enormous star."[6]

5. To give the student a better awareness of his or her particular technical discipline, for as Harvey Brace Lemon maintains, "Just as one can more fully appreciate a great telescope like the one on Mt. Palomar after the first simple glass of Galileo has been held in one's hand and looked through, so does the historical approach through the early stages of a great branch of science inspire interest, enthusiasm, and confidence in spite of the lack of any full understanding of details."[7]

In presenting these objectives, I emphasize that the series of assignments to follow will be at once practical and interesting. Most students agree with the forecast after they have completed the series.

Assignments

The series of assignments involves four areas, each of which will be briefly discussed.

1. Orientation—an introductory assignment to acquaint the student with the nature of scientific and technical writing, especially its steady awareness of audience
2. Rhetorical Modes—readings from scientific literature that illustrate effective use of various rhetorical modes
3. Style—readings from scientific literature that illustrate effective handling of sentence structure, vocabulary, and graphics
4. Disciplinary Readings—readings from scientific literature that demonstrate to the student how concepts basic to his or her technical specialty were handled by earlier writers

Technical writing textbooks commonly cite "audience analysis" as a central requirement. The phrase, of course, refers to the writer's obligation to ensure that material is carefully geared to the reader's background. In short, the technical writer must analyze the audience and be mindful of certain strictures. If, for example, the audience is comprised of lay readers, the technical writer must avoid technical terms that have not been defined adequately and provide sufficient background discussion so that readers are able to comprehend the thrust of the discussion.

The problem facing the teacher of technical writing is to convince students that scientists and those with technical expertise are quite capable of analyzing their audiences and gearing their discussions accordingly—especially when a lay audience is involved. Many scientists, of course, have written effectively for the general audience—Benjamin Franklin, to name one. But I have found that Michael Faraday's *On the Chemical History of a Candle* provides the best illustration of technical material prepared for a lay audience. Having read this remarkable series of six essays, students are immediately impressed with Faraday's ability to convey technical matter with clarity and precision. In turn, I have found that when they are asked to describe or define a physical object to a lay audience, these same students can apply their analysis of Faraday's work to their own writing tasks. Indeed, the results seem much more satisfactory than if I merely lecture on audience analysis and then ask the students to write something for a lay audience.[8]

Similarly, past scientific and technical literature can be very useful in helping students master the rhetorical modes common to technical writing, i.e., definition, classification and analysis, process description, and description of a mechanism. For best results, the instructor should lecture briefly on a given mode before handing out a selective reading list of materials for in-class discussion and analysis. A brief writing exercise for that mode concludes the assignment.

A rich and varied supply of reading material exists from which the instructor may select. The mode of definition, for example, is handled effectively in Robert Grosseteste's "On Light," Robert Boyle's "What Is an Element?", Charles Lyell's "Geology Defined," and William Gilbert's "A Preface to Electromagnetism." For classification and analysis, the following are useful: Aristotle's "On the Parts of Animals," Sir Julian Huxley's "The Size of Living Things," Marie Curie's "The Derivatives of Uranium," and John E. Teeple's "The Function of Chemical Engineering." For process description, I suggest Vitruvius' "On the Planning of Theatres," Agricola's "How Metals Are Produced," William Harvey's "On the Movement of the Heart and Blood in Animals," and Charles Darwin's "Natural Selection; or the Survival of the Fittest." And for description of a mechanism, I recommend Frontinus' "Aqueducts of Rome," Benjamin Franklin's "Description of a New Stove for Burning of Pitcoal," and Othmar Ammann's "Design of the Mackinac Bridge for Aerodynamic Stability."

Teachers of writing generally agree that one of the most effective ways to help students produce good writing is to have them *read* good writing—and as much of it as possible. The same holds true for technical writing; consequently, as an extension of the assigned readings in the preceding paragraph, I assign specific readings wholly to illustrate good technical writing style. Despite changes in our language (and the negative effect of translation in some cases), the best writing of the past remains an instructive model for students today. In addition, I have found not only that my students become more aware of good technical writing but that they also thoroughly enjoy reading such classics as John Wesley Powell's "Conquest of the Grand Canyon," John Muir's "Through the Foothills with a Flock of Sheep," and Dallas Lore Sharp's "Turtle Eggs for Aggassiz." And best of all, many students enhance their writing styles.

If an instructor chooses to rely on only one earlier writer, I can think of no better choice than Benjamin Franklin. As a complement to his many achievements, Franklin was committed to good writing. He once wrote, "It seems to me that there is scarce any Accomplishment more necessary to a Man of Sense, than that of *Writing well* in his Mother Tongue."[9] Furthermore, he believed that writing should be *"smooth, clear,* and *short"* and should have "not only the most expressive, but the plainest words."[10] Most significantly, Franklin practiced what he preached —as is evident in his *Experiments and Observations on Electricity* and his *Account of the New Invented Pennsylvanian Fire-Places.* In addition to being fine models of technical writing, both works contain admirable examples of graphic aids.

The final course exercise moves from the practical realm to the more intrinsic realm of appreciation—appreciation for one's technical discipline. Technical writing students frequently have difficulty relating to the history of their discipline. Most are so firmly entrenched in the immediate developments within their disciplines that they overlook the often fascinating and informative tradition of these disciplines. To address this problem, I assign reading units based on each student's discipline. The student is asked to read the material, conduct further research, and prepare an intermediate report (either for a layman or for someone majoring in that technical discipline) that concentrates upon writing as a mode of communication within the history of that discipline. Needless to say, the assignment is a real eye-opener for most students.

The reading units are not difficult to assemble. As one might assume, engineering and natural science materials are plentiful and accessible. A reading unit for engineering can include Aristotle, Archimedes, Vitruvius, and da Vinci. With a bit of research, an instructor can also compile useful reading units for chemistry, premed, biology, forestry, or any other technical discipline. To supplement each unit, I ask students to read Albert Einstein's "The Common Language of Science" and Thomas Van Osdall's "The Poetry of Science: A Cultural Force," and I am considering adding Lewis Thomas's *The Lives of a Cell* to the list.

The assignments outlined above are designed as complementary exercises in the traditional junior-senior technical writing course, but they would also be feasible in a junior college course. It should be evident that these assignments are appropriate for students from any of the disciplines usually represented in a technical writing course. Moreover, each assignment is flexible, and the instructor can adapt an assignment to reflect the personal reading preferences of students. The scheme has a certain organic quality in that scientific literature can be used throughout the course, or—

if time disallows the entire scheme—any one area or combination of areas can be employed successfully.

A nearly endless supply of primary materials is available to the instructor; almost every prominent scientist has had his or her writings collected. Many fine anthologies of scientific and technical literature are also readily accessible. Let me mention a few: *Through Engineering Eyes*, ed. Frank A. Grammer and James H. Pitman; *A Science Reader*, ed. Laurence V. Ryan; *A New Treasury of Science*, ed. Harlow Shapley, Samuel Rapport, and Helen Wright; *The Beginnings of Modern Science*, ed. Holmes Boynton; *New Worlds in Science*, ed. Harold Ward; *Science in Literature*, ed. Frederick Houk Law; *Great American Nature Writing*, ed. Joseph Wood Krutch; and *Great Adventures in Science*, ed. Helen Wright and Samuel Rapport.

Evaluation

Evaluating the success of these assignments depends largely on the instructor's sensitivity to the total development of each student. For example, an improved writing style is one easily detected sign of success. More permanent signs, such as maturing thought processes and a broader consciousness of one's discipline, may not be quite as evident. Be assured at least that these assignments generate specific benefits, and that in most cases these benefits are lasting:

1. The student is encouraged to write more clearly and precisely.
2. The student is encouraged to respond creatively to his or her technical discipline.
3. The student is encouraged to become a better thinker.
4. The instructor has at his or her disposal a workable, stimulating series of assignments.
5. Both the student and the instructor become more attuned to the role of written communication in science and technology as it has manifested itself from Aristotle to Einstein.

Because I believe these benefits are both important and appropriate in the technical writing classroom, I recommend that technical writing teachers experiment with exercises based upon the reading of past scientific literature.

Notes

1. Owen Barfield, *Saving the Appearances: A Study in Idolatry* (New York: Harcourt Brace & World, 1965), p. 11.

2. In developing this series of exercises, I was influenced significantly through correspondence with Herman Estrin of the New Jersey Institute of Technology. His article, "An Engineering Report Writing Course That Works," *Improving College and University Teaching* 26, no. 1 (Winter 1968): 28-31, and Walter J. Miller's "What Can the Technical Writer of the Past Teach the Technical Writer of Today?" *IRE Transactions on Engineering Writing and Speech* 4 (December 1961): 69-76, were also helpful.

3. Holmes Boynton, ed., *The Beginnings of Modern Science: Scientific Writings of the 16th, 17th, and 18th Centuries* (New York: Walter J. Black, 1948), p. xvii.

4. Boynton, p. xviii.

5. I have borrowed the phrase from Thomas Van Osdall, "The Poetry of Science: A Cultural Force," *Journal of Chemical Education* 50, no. 3 (March 1973): 174-75.

6. Frederick Houk Law, *Science in Literature* (New York: Harper & Brothers, 1929), p. xv.

7. Harvey Brace Lemon, Foreword, *The Beginnings of Modern Science* ed. Holmes Boynton, p. viii.

8. Another excellent example of a man of technical expertise writing to a specific audience is Arthur M. Wellington's monumental *Economic Theory of the Location of Railways*, 6th ed. rev. (New York: John Wiley & Sons, 1906). Originally published in 1887, the book accommodates three levels of audience: layman, semi-expert, and expert.

9. L.W. Labaree and W.J. Bell, Jr., eds., *The Papers of Benjamin Franklin*, 18 vols. to date (New Haven: Yale University Press, 1959-), 1 : 328.

10. Labaree and Bell, 1 : 329.

9 *Scientific American* in the Technical Writing Course

Wayne A. Losano
University of Florida

Good examples are useful in the technical writing class but are not always adequately provided in technical writing textbooks. *Scientific American* can be used as an excellent and inexpensive source of examples to teach audience adaptation, organization, technical style, definitions and descriptions, graphics, and formal report elements. This paper illustrates how that can be done by referring to specific articles from *Scientific American*.

Regardless of the technical writing text one selects, the paucity of examples of good technical or scientific writing remains a major problem. I have found that articles in *Scientific American* fill this gap nicely. The magazine, readily available and inexpensive, provides concrete illustrations of most facets of technical writing and is especially useful to technical writing teachers without industrial experience from which to draw material for classroom use.

I have used *Scientific American* on three levels: in a basic technical writing class (undergraduates only, predominantly sophomores and juniors of such majors as engineering, agriculture, and business); in an advanced report writing class (seniors and graduate students, usually premed, prelaw, chemistry, physics, and journalism majors); and in a graduate course for English majors: Teaching Business and Technical Communication. The magazine would work equally well in other specialized courses such as Writing for Publication and Technical and Scientific Journalism.

Objectives

In general, I use *Scientific American* articles to ensure that students are familiar with examples of competent technical writing

94

and can recognize the effective implementation of the various technical writing tenets covered in class. More specifically, articles in *Scientific American* can be used to acquaint students with such aspects of style as concreteness and economy, with general organizational patterns, with the proper function of graphic aids, and with the role of introductory and terminal sections.

One major, and reasonably definable, objective is to assess the abilities of students to select those aspects of professional writing that contribute to the overall success of a technical or scientific manuscript. Coherent organization and paragraph development, effective style, wise use of illustrations—all are readily evident in *Scientific American* articles, and near the end of a technical writing course students can be expected to do some thorough analysis. Generally, when I use this assignment, I give the following directions: "Analyze the following article, pointing out what might escape the casual, untrained reader. Discuss—and illustrate with concrete examples from the article—the article's organizational pattern, its style, the presentation of technical data in prose and in visual form, the function of various sections, and other factors that contribute to the article's success as technical or scientific writing." Students have produced some remarkably insightful analyses, and have indeed pointed out much that would escape most untrained readers. I require that these analyses be presented in a professional manner: well written, clearly organized, and thoroughly illustrated. Students are cautioned to approach an article by topic rather than to attempt a paragraph-by-paragraph summary/analysis, as might be their tendency. The three or four students submitting the best analyses then function as a panel to lead a class discussion of the article.

The meeting of this particular objective—the ability of students to detect and clearly define factors that make for a successful technical article—must, I think, be accompanied with a warning. In most on-job situations, pressures of time, inadequacies of data, or other factors may well preclude the achievement of genuinely high standards of technical writing. Often, we should warn students, "quick and dirty" writing jobs must suffice. Thus, it is perhaps best to blend discussion of articles from *Scientific American* with discussion of readily available but poorer models. Nonetheless, good writing is the goal, if seldom attained in other than published articles, and we should hold up the better models for our students.

Implementation

I begin by requiring each student to buy a copy of the current issue of *Scientific American.* Then I use it from time to time by tying various aspects of specific articles into my regular lectures and discussions, referring to the articles to reinforce the illustrations provided by the course text. In early discussions of audience levels, for example, I use *Scientific American* articles to illustrate material suitable for the intelligent layperson but generally too theoretical for technicians and too basic for experts. Several exercises are possible at this time: articles can be further simplified for the general or mass audience, definitions and clarifying illustrations can be removed to make the articles better suited to a higher-level readership, and students with the necessary technical background may attempt to revise articles for the more practical technician.

Numerous other aspects of technical writing can be taught in this way. Indeed, I use *Scientific American* to teach each of the following topics.

Organization

Scientific American articles lend themselves nicely to discussion and illustration of methods of paragraph development and organization. An easy exercise is to ask students to read through an article and insert appropriate headings. The magazine does not provide headings but does indicate new sections with space, so students are given some direction for first-level headings. Close reading, however, is required to come up with useful headings, and subheadings present an even greater challenge. Devising headings leads to a discussion of methods of organization and outlining, and various examples can be found in each issue of *Scientific American.* Historical or chronological development is used in "The Fabrication of Microelectronic Circuits" in the September 1977 issue; comparison and contrast is used in "Microelectronic Memories" in the same issue; illustration and example can be found in "Kangaroos" in the August 1977 issue, and so on. Any rhetoric text provides the instructor with information on the various methods of development and organization, and *Scientific American* can serve to illustrate most methods.

Style

The majority of *Scientific American* articles are well written, so there is little opportunity for the sort of sentence revision exercises found in most texts. Nonetheless, some stylistic revision is often possible, usually on a more sophisticated level than that demanded by textbook exercises. For example, in "Light-Wave Communications" in the August 1977 issue, the sentence "The first high-transparency fibers were made by the Corning Glass Works out of a material whose principal component was silicon dioxide" is open to the charge of wordiness and lends itself to a discussion of active versus passive construction (and in this case, the passive wins). An adequate illustration of sentence balance, from the same article, is "Just as there are two kinds of light source for light-wave communications, so there are two types of detector in service." Class discussion can center on stylistic economy and concreteness as well as parallelism, coordination, and subordination. I have also found it useful to compare the often "leisurely" style of *Scientific American* articles with that of "hard core" examples of technical writing I have gathered from various companies.

Definition and Description

Most technical writing texts provide at least a chapter on formal sentence definitions, expanded definitions, and descriptions of mechanisms and processes. Reading *Scientific American* allows students to observe these technical writing strategies in action.

Random sentence definitions from recent issues include "Capacitance is a measure of the electric field surrounding a conductor," "Inductance represents the energy stored in the magnetic field set up by an electric current," and "Tempering is a practice that reduces the brittleness induced by quenching." I asked students to carry on from the first definition, defining *electrical field* and *conductor* for the layperson, and we found the use of the word *represents* in the second definition somewhat tricky, thus requiring clarification.

An expanded definition appears in "The Structure and Function of Histocompatibility Antigens" (October 1977): "Histocompatibility antigens are defined as any molecules that differ from individual to individual and are not recognized in graft rejections."

This formal definition is developed, with illustrations, for four paragraphs. A shorter example is found in "X-Ray Stars in Globular Clusters" in the same issue: "Globular clusters are self-gravitating systems consisting on the average of about 100,000 stars. Some 130 of them are known at present and perhaps 70 more may be hidden from us by intervening dust clouds. They populate the galactic 'halo': the spherical region that is bisected by the disk of the galaxy. All told the globulal clusters consist of some 10 million stars." Adequate examples can be found in nearly all articles.

The article on histocompatibility antigens also offers several good examples of technical description, including a nicely complex process description of the sequencing of the H-2 heavy chain, complete with visual aid. The description of the process of analyzing the amino acid sequence of H-2 heavy chains in the October 1977 issue lends itself to conversion to basic instructions, another useful exercise. Samples of descriptions of processes or mechanisms can be found in most articles in *Scientific American.*

Graphics

Articles in *Scientific American* are well illustrated, and technical writing teachers will find a variety of useful visual aids to discuss. Briefly, "Kangaroos" (August 1977) offers photographs, diagrams, sketches, maps, line graphs, and bar charts, all of which are effectively tied in with the text. Other articles are similarly fruitful.

Formal Report Elements

Scientific American articles are especially useful to illustrate the form and function of various elements of the formal report. Some article titles can be discussed; e.g., the too-general "Kangaroos" versus the specific "The Flow of Heat from the Earth's Interior," in the August 1977 issue, and the romantic "Fundamental Particles with Charm" versus "The Structure and Function of Histocompatibility Antigens" in the October 1977 issue.

Introductory paragraphs or sections in most of the articles illustrate the inverted pyramid approach suggested by many textbooks. "Side-Looking Airborne Radar" (October 1977) is a good example. Paragraph one is a general, historical survey; paragraphs two and three move through efforts to develop efficient aerial photogrammetry and the limitations of these efforts; paragraph four discusses the newest, most efficient method; paragraph five expands on this method and outlines the article to follow.

Thus, both functions of formal introductions—to introduce the subject matter and to introduce the article—are illustrated.

Terminal sections are similarly well illustrated. "Kangaroos" has a basic summary conclusion, reversing the inverted pyramid introduction and reemphasizing the main points of the article concerning the kangaroo's amazing adaptation. The concluding paragraph is also a fine example of topic sentence development, growing from the initial sentence: "The picture of the kangaroos that has now been developed is one of a group of mammals that are not at all primitive but have adapted and radiated rapidly in response to a new and changing environment." Interestingly, the final sentence of the paragraph points disturbingly to a contrasting phenomenon—the disappearance of the kangaroo's smaller relatives, wallabies and rat kangaroos, resulting from the extension of the arid environment in which kangaroos thrive—thus further highlighting the kangaroo's splendid adaptation.

The conclusion of "Fundamental Particles with Charm" is more speculative, pointing out avenues for further investigation: "Further experiments will be needed to confirm this discovery. . . ." Various other articles provide other useful examples.

Evaluation

The variety of uses one can find for *Scientific American* articles makes clear definition of evaluative methods a bit difficult. In general, students are evaluated on their abilities to define and illustrate tenets of effective technical and scientific writing. Specifically, they can be asked to analyze segments of articles for function and effectiveness and to suggest and defend modifications. Although I use the articles primarily for class discussion, I do require written critiques of entire articles. I have also had students prepare informative abstracts of articles, and I have had them attempt the letter of query that might entice an editor to publish a particular article. I suspect most teachers of technical writing could find additional uses for *Scientific American* in their classes; the specific uses would then determine the method of assessment.

10 Cut and Paste: Preparing for On-the-Job Writing

Gordon E. Coggshall
Technical Communications Department,
Perkin-Elmer Corporation

As all experienced technical writers know, writing is frequently done under intense pressures by writers who begin their task with a collection of materials from which a new document must be built quickly and effectively. Cut-and-paste exercises develop this skill and reinforce other types of instruction. Four cut-and-paste assignments are briefly described, and for one the complete assignment and an example of its successful completion are given.

The technical writing classroom is most successful when it approximates the conditions students will find on the job. There, efficient writers often insert previously written materials in new reports or, when revising, cut apart a completed draft and physically reorganize it. Both activities are examples of *cut-and-paste*. Because most college students have never needed that technique in other courses, they tend to be scornful of it as a half-measure, as uncreative, even as cheating. But if the instructor presents a series of progressively more difficult cut-and-paste exercises, each in a context that students might actually find in their careers, students soon learn to respect the technique. More important, the technique challenges, then reinforces, student abilities to organize material according to a logical format, to revise, and to edit.

In technical writing classes at Ithaca College I teach juniors and seniors from a variety of professional disciplines: business, health-services administration, physical therapy, music, communications management, education, recreation. Most of these students view themselves as preprofessionals, and they are generally experienced and aware of what it means to work in their professions. Their composition skills are generally above average, but they have not yet learned efficient methods of producing competent technical writing within the intense time pressures they will frequently find on the job. Thus these students are ready for pragmatic approaches to writing, one of which is cut-and-paste.

The Assignments

I assign a series of four cut-and-paste assignments, the first two to be completed in class and the last two as homework. The four assignments, each more difficult than the one before, are spaced throughout the semester, and each follows the study of a particular format: proposal, progress report, journal article, formal report. Thus the cut-and-paste assignments provide additional work with particular formats, reinforcing them.

The first assignment asks students to unscramble an informal proposal and to put it in the form of a business letter. I distribute copies of a scrambled proposal, on the last page of which is a list of headings. Working in pairs, students devise an outline that suits the format, cut the proposal apart, and tape the segments in a logical order. When they have finished reorganizing the parts, they are instructed to check the new draft for proper format, to add useful headings, and to edit for clarity. Finally, they are asked to attach instructions to a typist, instructions that would ensure an acceptable page layout.

The second assignment—given in full, together with one student's response, at the end of this paper—is more difficult because more than reorganization is required and because no headings are given as clues to organization. For this assignment, given soon after periodic-report format has been covered in class, I prepare a double-spaced typed transcript of a loosely dictated progress report. The dictation skips around, as if its author had been spilling out random memories of work done during the report period (three months in the middle of a two-year project). Moreover, the sentences are choppy, the use of jargon is troublesome, and considerable useless information is given. The student is instructed to choose an appropriate progress-report format, to make an outline, to cut and paste the transcript, to revise and edit it heavily, to note where visuals or supplementary materials are to be inserted, and once more to give the typist instructions.

The third assignment, given soon after publishing articles in professional journals has been covered in class, provides a jumble of more or less raw materials for a journal article. The jumble includes correspondence, excerpts from journal articles on a similar topic, graphs and charts, raw data, excerpts from a project feasibility study, and other assorted pieces of writing. Students are asked to take the material home and fashion from this miscellaneous collection, almost certainly a collection similar to one they will discover on their desks at work several years hence, the draft of an article for a professional/technical journal. By cutting apart, reorganizing, taping, revising, editing, and writing instructions to a typist, students once again fashion technical pieces under pressure of time.

The final cut-and-paste assignment is similar to the third but more extensive. I provide a packet of about fifteen pages of letters, newspaper articles, proposals, progress reports, tables of raw data, graphs and charts of digested data—even several pages of computer printout. I also place on library reserve a book-length file of supplementary reading. I instruct students to come up with a draft of a final project report, complete with suggested appendixes. Among the material I hand to students is material that is not properly part of a final report (uninterpreted data, part of a journal article on a subject unrelated to the project). Since each piece of material is by a different author, students must not only reorganize and revise but must also edit for tone and voice.

Evaluation

If students have completed the cut-and-paste assignments successfully, they will have demonstrated the following skills:

1. An understanding of the concept of format generally and the ability to use a specific format (proposal, progress report, journal article, and formal report)

2. The ability to outline so that a subject is covered logically, exhaustively, and in a way appropriate to the format

3. The ability to visualize page layout and to communicate that layout to someone else, theoretically a typist but in this case the instructor

4. The use of headings, titles, and subtitles

5. The use of visuals

6. The use of cross-references

7. The ability to revise and edit, especially for precision and clarity, for uniformity of voice and tone, and for suitable levels of technical language

8. The ability to choose appropriate appendix material

Because the cut-and-paste assignments can be completed in a number of successful ways, no single answer key can be made against which to compare each student's work. Instead, I prepare a checklist that parallels the list of skills applicable to the assignment under consideration and assign points in proportion to the importance that I place on each skill. The checklist that accompanies the second assignment (the progress report) is given below along with the maximum number of points that can be earned for each item, based on a total score of 100 points.

> The finished progress report uses an appropriate format (10 points).
>
> The report has a logical organization (15 points).
>
> The report is as complete as the raw materials allowed (10 points).
>
> The report and the instructions to the typist reflect the author's attention to page layout (5 points).
>
> Headings, titles, and subtitles are appropriate and informative (15 points).
>
> Visuals are clear (10 points).
>
> Visuals are properly placed and cross-referenced (15 points).
>
> The report shows proof of good revising and editing skills (20 points).

Evaluation is necessarily subjective; however, the checklists serve as a guide for both the teacher and the students, and I have found that both poorly done and well-executed assignments are clearly distinguishable.

Second Cut-and-Paste Assignment: Unedited Transcript and Student Response

Below is the second cut-and-paste assignment. Following the unedited transcript is an example of its successful transformation to a progress report. Notice that the content is likely to be unfamiliar to the class, and the technical language of biology presents an opportunity for the instructor to emphasize the need to avoid jargon and to provide definitions suitable to the audience. Notice, too, that in the completed report material from the transcript was eliminated when it was unnecessary or digressive. Not every cut-and-paste exercise need have problems of jargon or digression, but such exercises are ideal for checking the student-editor's skills in managing these complex problems.

Unedited Transcript of a Progress Report

Directions: Imagine that what follows is a word-for-word, unedited transcript of the dictation by a project director (Irving H. Washington)
detailing three months' work on a project testing the growing of fish in
closed tanks as potential food for interplanetary space travel.   Decide on
a progress-report format, make an outline, formulate subtitles and headings,
and cut and paste the transcript into a progress report.   Then edit and
revise.   Finally, add whatever instructions you think are necessary for a
typist to produce a finished copy.

This is a report of three months on the project to study the

use of fish for food aboard interplanetary space travel.   Work is being

performed under contract #304-3054 from NASA.   Total amount of funds:

104,000 dollars U.S.   Total expenditures to date: $62,400.   Breakdown:

$36,000 oaboratory equipment, $16,200 salary, wages, and benefits,

$8,000 overhead, $1,200 travel, and $1,000 consultant's fee.   Duration

of project: 1 Jan. 1976 through 31 December 1977 (two years).   Period

of progress report: 1 July 1977 through 30 September 1977 (three months).

Progress is on schedule.   Prior to 1 July all work on choosing the

optimal species was completed.   We had also completed construction of

tanks and purchase of test equipment.   During the first week in July

data on weight gain and relative wieght of fist parts were assembled,

regressed, and is presented in the table one (attached).   From these

data, Tilepia mossambica Peters, a member of the Cichlidae family of

fish, was chosen as most promising to be grown aboard a spacecraft.

The criteria for making that selection (which we made on 15 July 1977)

are given as a list (attached).

During the remainder of July and throughout August, experiments

to see the limits of environmental plasticity of Tilqpia were carried

our.   Oh, by the way, environmental plasticity means the range of en-

vironmental change--temperature, pressure, food supply, etc.--that a

species can withstand without adverse health effects.   Profect members

felt the work went well, although some re-design of equipment was necessary.

For example, the means of keeping tank water at a given temperature had to be modified to keep the temperature more controllable. Similarly, oxygenation equipment had to be modified to better distribut oxygen throughout the tanks. The results of these fluctuation experiments are given in table two (attached) and show not the <u>extreme</u> limits of plasticity but rather the limits under which the Tilapia grow sufficiently fast and multiply sufficiently rapidly to provide food for the necessary animal-protein portion of the human diet.

Oh, and list I forget, for you who are new readers or who are unfamiliar with the project, here's a project overview. NASA funds were granted to study fish culture aboard spacecraft. Humans need forty percent of their protein requirement to be animal protein (contrary to the line of bull from some crazy vegetarians), that is, protein from heterotrophic organisms. Fish have been suggested as most promising because of their efficient generation of animal protein and because of their ability to withstand a wide variety of envoronmental variations, including the rapid accelerations accompanying space flight. Our project, entitled "Use of Fish as Animal Components of a Closed Ecological System," will choose the best fish, study the limits of its endurance, measure and record its food-prodicing capability, and determine (roughly) its contribution to negentropy (from negative entropy, the amount of unretrievable energy lost from the closed system).

Now, on to the work of the last period (1 October to end of project). We have left only two things. First, we will be coming up with values of negentropy. These calculations (based on formulae determined by N. V. Mironova, a brillian Russian--Soviet--physicist) will be made by computer, based on a study of the informational complexity of the various proteins in the parts of the fish. Calculations will be imprecise, but we will be in a good position, after the first round of calculations, to suggest ways in which calculations may be made better (in preparation for an addendum proposal seeking monies to make a second round of negentropy calculations).

Then in December 1977 we will be writing up the repot of our work, seeking to also publish four articles in scientific periodicals. We will see that our final report is sent to NTIS (National Technical Information Service) for insertion in it computer information-retrieval system. Also in December, along with the report, we will prepare our addendum proposal to continue the portion of our work that has to do with negentropy calculations.

Oh--I almost forgot--In September this year, we also began to study the proteins under electron microscopes in preparation for our calulations of protein complexity for our negentropy calculatons. A consultant was hired to design the study of the photomicrographs and the program for computer study of data taken from the photomicrographs. Photomicrographs are pictures taken directly through the lenses of a microscope. The consultant was S. G. Ratner, an expert in electromicroscopy (computer applications thereof) from NIH (National Institutes of Health), on leave. He spent one week in helping us design our data-use techniques regrading Protein study.

Other things . . . some benchmarks of the entire project: 1 july 1976 finished fishtanks and genral laboratury equipment. 1 August 1976 collected species of fish to study (except white amur, which had to be imported from Lithuania and arrived a month into the experiment. 15 November 1976: ended first round of comparative growth of fish; also dropped the notion of testing limits of gravitational (accelerative) stress because the cost of centrifuge equipment and fish-tank modifications were extravagant.

And I forgot: In July (end of July 1977) we studies weight gain (growth) of Tilapias. See table five (attached. Also, in a separate set of tanks we tested the ability of several species (including tilapia) to ingest various kinds of garbage, including human offal. This is important, since fish in interplanetary flight would act as a kind of lvving garbage disposal system. Data on these tanks are not available at the time of this report. Both these sets of experiments went on side-by-side with the environmental plasticity experiments of July end.

And in the last two weeks of August (also side-by-side with plasticity experiements) we did fertility experiments; see table 8, attached.

TABLE L. Relative Weights of Various Parts of the Body of
Fishes (in % of Live Weight)*

Species	Sex	# of Spec.	Weight of fish in g.	Head	Viscera	Fins& bones	Meat& Skin	Scales	Total	Loss of blood & moisture
Tilapia	♀	1	130.35	19.64	11.43	17.72	43.54	3.49	95.82	4.18
"	♀	1	62.9	18.84	8.82	16.38	46.02	4.06	94.12	5.88
"	♀	1	126.2	18.34	4.91	9.51	52.02	4.52	89.30	10.70
"	♀	1	66.85	18.48	6.73	10.92	49.53	4.26	89.98	10.02
"	juv	5	18.7	20.05	8.13	11.76	.53	.48	93.42	6.58
Carp	?	?	428	16.0	12.1	14.1	52.4	4.5	100.0	
White amur.	?	?	794	13.4	9.4	12.4	61.2	3.6	100.0	
Silver carp	?	?	449	17.0	10.6	12.4	55.6	4.4	100.0	
Big head	?	?	703	23.2	7.6	14.6	15.1	3.5	100.0	

*The weight of blood and moisture lost during dissection is not distinguished in the data cited from the literature. In our results they are shown in a separate column. If this amount is added to the weight of meat, the results obtained are identical to those characteristic of the carp, while amur, silver carp and bighead.

1. High ecological plasticity.

2. Omnivorous and mainly phytophagous feeding habits. Ability to feed on algae, green mass of higher plants, and animals.

3. A supposed ability to feed on <u>Chlorella</u>.

4. A technically convenient optimum temperature of existence of between 20 and 28°, not requiring special thermoregulatory devices for its matintenance.

5. Resistance to changes in temperature over a wide range (from +8-10° to nearly 40°).

6. Absence of denand of seasonal changes in temperature conditions.

7. Absence of seasonal cyclic pattern of growth and reproduction (with consequent regular increase in biomass).

8. Early maturation, frequent spawning all the year round, caring for offspring, rapid increase in population of biomass.

9. High rate of growth at an early age and attainment of suitable dimensions for consumption as food within a few months.

10. Ability to exist in a high stocking density without appreciable influence on growth and propagation, reported in the literature (12).

TABLE 2. Fluctuations in Conditions of Existence of
Tilapia in Experimental Aquaria

Indices	Minimum	Maximum
Water temperature	17.9	37.6
Mean water temperature for the month.	20.7	27.3
Fluctuations in water temperature during month. .	1.0	17.5
Fluctuations in water temperature during day. . .	0.0	17.5
pH.	6.0	6.8
Oxygen concentration in water, mg/liter	.85	8.37
Oxygen saturation of water, %	9.6	94.5
B. O. D. (permanganate)	11.8	17.3
B. O. D. (bichromate)	23.2	67.2

TABLE 5. Growth of Tilapias per Month as Percentages
of their Weight at the age of 12 months

Aquarium No.	No. of speci- mens	Months											
		0-2	2-3	3-4	4-5	5-6	6-7	7-8	8-9	9-10	10-11	11-12
1L4	6	2.1	5.1	8.9	8.2	9.1	14.2	9.7	19.4	10.6	4.8	7.2
2L4	6	2.2	6.1	7.2	5.2	9.3	13.2	12.6	13.7	14.1	5.8	10.6
L10	6	1.7	5.5	6.7	9.3	12.3	14.4	14.2	11.2	9.5	3.7	11.0
L10A	6	2.0	8.6	10.6	11.0	12.6	11.0	11.4	3.9	10.3	9.5	9.7
L53	12	1.9	6.3	15.0	11.2	10.9	12.0	10.7	7.5	9.6	8.0	6.4
Mean		2.0	6.3	9.7	8.9	10.3	12.9	11.7	11.1	10.8	6.1	9.0

TABLE 8. Reproduction in Six Female Tilapias from
August 18, 1976 to August 10, 1977

Indices	Number of females					
	5/10	10	8_1	8_3	8_2	4
Number of spawnings	13	8	9	9	6	3
Age at first spawning, months . .	6	6	5	6	5	7
Weight, g	11	11	10	9	2.5	13.5
Mean interval between spawnings, days	25	41	44	38	71	60
Length, mm.	88	90	82	--	65	90
t^o min at time of spawning. . . .	21.6	22.2	23.5	22.5	22.2	22.5
t^o max at time of spawning. . . .	26.5	25.9	26.4	27.7	25.1	28.7
t^o mean during incubation	24.1	24.9	25.0	24.1	24.0	26.1
Mean incubation time, days. . . .	11	11	11	9	9	--
Fertility	90,77 30	25,60 209, 20,41	31,15 184,70	102,97 149	70,40	--
Number of larvae reared..	90,77	60, 209	--	--	--	--
Number surviving six months . . .	42,77	14, 204	--	--	--	--

Edited Version Prepared by a Student

[typist: single-space ¶'s;
double-space headings
and between ¶'s; triple-
space before and
after figures]

Progress Report:
Use of Fish as Animal Components of a
Closed Ecological System

~full caps~   Irving H. Washington, Project Director

(Nasa) Contract 304-3054   Project Funds: $104,000
Project Duration: 1 January 1976 through   Expenditures to Date: $62,400
                  31 December 1977
Report Period: 1 July 1977 through
               30 September 1977

Project Overview   ¶ NASA funds

were granted to study fish culture aboard spacecraft. Humans need forty

percent of their protein requirement to be animal protein, ~(contrary~

~to the line of bull from some crazy vegetarians),~ that is, protein from

                              are
heterotrophic organisms. Fish, ~have been suggested as~ most promising

because of their efficient generation of animal protein and because of

                                                         i
their ability to withstand a wide variety of environmental variations,

                                                          This
including the rapid accelerations accompanying space flight. ~Our~ project,

entitled "Use of Fish as Animal Components of a Closed Ecological System,"
                   species of
will choose the best, fish, study the limits of its endurance, measure

                              u
and record its food-prod/cing capability, and determine ~(roughly)~ its

contribution to negentropy (~from~ negative entropy, the amount of
                         irretrievably
~unretrievable~ energy lost, from the closed system).

Work Prior to Report Period

     ~Other things . . . some benchmarks of the entire project:~ On
                                                                  1 July
       technicians constructing   By   researchers
1976, finished, fishtanks and general laboratory equipment. 1 August 1976,
had   all
, collected, species of fish to study (except white amur, which had to

be imported from Lithuania and arrived a month into the experiment)
                 marked the   of the   studies measuring
15 November 1976, ended, first round of, comparative growth of fish. On the same

[date project directors] ∧also dropped the notion of testing limits of gravitational (accelerative) stress because the costs ∧of *[both]* centrifuge equipment and ∧fish-tank *[of]* modifications were extravagant. *[typist: align type]* *(Tilapia)* Prior to 1 July ∧*[1977]* all work on choosing the optimal species ∧was completed. ~~We had~~ also completed ∧*[was]* construction of tanks and purchase of test equipment.

Progress During Report Period *[Throughout]*

~~During~~∧ the first week in July ∧ data on weight gain and relative w∧eight *[h]* of fis∧ parts were assembled∧ *[and]* regressed, and∧ *[are]* ~~is~~ presented in the ṯable ~~one.~~ *[1]* ~~(attached)~~∧

    TABLE *1.* ~~I.~~ Relative Weights of Various Parts of the Body of
            Fishes (in % of Live Weight)∧[a]

[reverse row and column headings so that table is vertical rather than horizontal]

Species	Sex	# of Spec.	Weight of fish in g.	Head	Viscera	Fins& bones	Meat& Skin	Scales	Total	Loss of blood & moisture
Tilapia	♀	1	130.35	19.64	11.43	17.72	43.54	3.49	95.82	4.18
"	♀	1	62.9	18.84	8.82	16.38	46.02	4.06	94.12	5.88
"	♀	1	126.2	18.34	4.91	9.51	52.02	4.52	89.30	10.70
"	♀	1	66.85	18.48	6.73	10.92	49.53	4.26	89.98	10.02
"	juv	5	18.7	20.05	8.13	11.76	.53	.48	93.42	6.58
Carp	?	?	428	16.0	12.1	14.1	52.4	4.5	100.0	
White amur.	?	?	794	13.4	9.4	12.4	61.2	3.6	100.0	
Silver carp	?	?	449	17.0	10.6	12.4	55.6	4.4	100.0	
Big head	?	?	703	23.2	7.6	14.6	15.1	3.5	100.0	

a∧The weight of blood and moisture lost during dissection is not distinguished in the data cited from the literature. In our results they are shown
in a separate column. If this amount is added to the weight of meat, the
results obtained are identical to those characteristic of the carp, whi~~l~~e∧
amur, silver carp∧ and bighead.

[no new ¶]
←————————————————————————————————From these
data, <u>Tilepia mossambica Peters</u>, a member of the Cichlidae family of
[species]
fish, was chosen as most promising∧ to be grown aboard a spacecraft.

The criteria for making that selection (which we made on 15 July 1977)
are given ~~as a list (attached)~~∧ *below.*

[typist: increase right margin as shown here.]

    1. High ecological plasticity.

    2. Omnivorous and mainly phytophagous feeding habits. | Ability
to feed on algae, green mass of higher plants, and animals.

3. A supposed abilty to feed on <u>Chlorella</u>.

4. A technically convenient optimum temperature of existence of between 20 and 28°, not requiring special thermoregulatory devices for its maintenance.

5. Resistance to changes in temperature over a wide range (from +8-10° to nearly 40°).

6. Absence of demand of~~ ~~ for seasonal changes in temperature conditions.

7. Absence of seasonal cyclic pattern of growth and reproduction (with consequent regular increase in biomass).

8. Early maturation, frequent spawning all the year round, caring for offspring, rapid increase in population of biomass.

9. High rate of growth at an early age and attainment of suitable dimensions for consumption as food within a few months.

10. Ability to exist in a high stocking-density without appreciable influence on growth and propagation. ~~reported in the literature (12).~~

~~And I forgot:~~ In July ~~(end of July 1977)~~ also we studied weight gain ~~(growth)~~ of Tilapias (see table ~~five (attached).~~ 2 below). Also, in a separate set of tanks we tested the ability of several species (including tilapia) to ingest various kinds of garbage, including human offal. This ability is important, since fish in interplanetary flight would act as a kind of living garbage disposal system. Data on these ~~tanks~~ tests are not yet available.

TABLE 2. Growth of Tilapias per Month as Percentages of their Weight at the age of 12 months

Aquarium No.	No. of speci- mens	Months										
		0-2	2-3	3-4	4-5	5-6	6-7	7-8	8-9	9-10	10-11	11-12
1L4	6	2.1	5.1	8.9	8.2	9.1	14.2	9.7	19.4	10.6	4.8	7.2
2L4	6	2.2	6.1	7.2	5.2	9.3	13.2	12.6	13.7	14.1	5.8	10.6
L10	6	1.7	5.5	6.7	9.3	12.3	14.4	14.2	11.2	9.5	3.7	11.0
L10A	6	2.0	8.6	10.6	11.0	12.6	11.0	11.4	3.9	10.3	9.5	9.7
L53	12	1.9	6.3	15.0	11.2	10.9	12.0	10.7	7.5	9.6	8.0	6.4
Mean		2.0	6.3	9.7	8.9	10.3	12.9	11.7	11.1	10.8	6.1	9.0

~~at the time of this report.~~ Both these sets of experiments went on side-by-side with ~~the~~ environmental plasticity experiments. ~~of July end.~~

⌒And in the last two weeks of August (also side-by-side with ^Continuing plasticity experiements) we did fertility experiments ῤ (see table 8, ~~attached~~ below.).

TABLE 8. Reproduction in Six Female Tilapias from
August 18, 1976 to August 10, 1977

Indices	Number of females					
	5/10	10	8_1	8_3	8_2	4
Number of spawnings	13	8	9	9	6	3
Age at first spawning, months . .	6	6	5	6	5	7
Weight, g	11	11	10	9	2.5	13.5
Mean interval between						
spawnings, days	25	41	44	38	71	60
Length, mm.	88	90	82	--	65	90
t^o min at time of spawning. . . .	21.6	22.2	23.5	22.5	22.2	22.5
t^o max at time of spawning. . . .	26.5	25.9	26.4	27.7	25.1	28.7
t^o mean during incubation	24.1	24.9	25.0	24.1	24.0	26.1
Mean incubation time, days. . . .	11	11	11	9	9	--
Fertility	90,77 30	25,60 209, 20,41	31,15 184,70	102,97 149	70,40	--
Number of larvae reared	90,77	60, 209	--	--	--	--
Number surviving six months . . .	42,77	14, 204	--	--	--	--

During the remainder of July and throughout August, experiments
to see the limits of environmental plasticity of Tilapia were carried
out. ~~Oh, by the way,~~ (environmental plasticity means the range of environmental change--temperature, pressure, food supply, etc.--that a
species can withstand without adverse health effects). Project members
felt ^that the work went well, although some re-design of equipment was necessary.
(For example, the means of keeping tank water at a given temperature
had to be modified to keep the temperature more controllable; Similarly, oxygenation equipment had to be modified to better distribute
oxygen throughout the tanks.) The results of these fluctuation experiments
are given in table two, (~~attached~~ ^below,) and show not the extreme limits of

TABLE ~~2~~ 4. Fluctuations in Conditions of Existence of
Tilapia in Experimental Aquaria

Indices	Minimum	Maximum
Water temperature	17.9	37.6
Mean water temperature for the month.	20.7	27.3
Fluctuations in water temperature during month. .	1.0	17.5
Fluctuations in water temperature during day. . .	0.0	17.5
pH.	6.0	6.8
Oxygen concentration in water, mg/liter	.85	8.37
Oxygen saturation of water, %	9.6	94.5
B. O. D. (permanganate)	11.8	17.3
B. O. D. (bichromate)	23.2	67.2

plasticity but rather the limits under which the Tilapia grow sufficiently

fast and multiply sufficiently rapidly to provide food for the necessary

animal-protein portion of the human diet.

¶ ~~Oh--I almost forgot--~~ In September ~~this year~~ we also began to study

the proteins under electron microscopes in preparation for our calulations

of protein complexity for our negentropy study. ~~calculatons~~ We hired a ∧consultant

~~was hired~~ to design the study of the photomicrographs and the program

for computer study of data taken from the photomicrographs. ~~Photomicro-~~

~~graphs are pictures taken directly through the lenses of a microscope.~~ ¶

The consultant was S. G. Ratner, an expert in electromicroscopy (computer

applications) ~~thereof~~ from (NIH) (National Institutes of Health), on leave.

He spent one week in helping us design our∧ data-preparation and data-use techniques ~~regarding~~ for

protein study.

Work Remaining

~~We have left~~ only two things remain to be completed on this project. First, we will be coming up with values

of negentropy. These calculations (based on formulae determined by

N. V. Mironova, a ~~brillian Russian~~ Soviet physicist) will be made by
computer, based on a study of the informational complexity of the various
proteins in the parts of the fish. Calculations will be imprecise, but
we will be in a good position, after the first round of calculations,
to suggest ways in which calculations may be made better, ~~(in preparation~~
~~for an addendum proposal seeking monies to make a second round of~~
~~negentropy calculations)~~

Second,
~~Then,~~ in December 1977 we will be writing up the report of our work,
seeking to also publish four articles in scientific periodicals. We
will see that our final report is sent to ~~NTIS~~ (National Technical
(NTIS)
Information Service) for insertion in it computer information-retrieval
system. Also in December, ~~along with the report~~ we will prepare our
addendum proposal to continue the portion of our work that has to do
with negentropy calculations.

11 Using Technical Information Resources

Maurita Peterson Holland and Leslie Ann Olsen
The University of Michigan

Many students are limited in their ability to do technical work or to write about it because they do not know how to research technical subjects in the library. Two methods of providing this instruction were developed jointly by a technical writing instructor and a technical librarian and are presented here.

The technical writer's credibility rests upon adequate knowledge of the technical subject and at times directly upon the ability to conduct a thorough search of the existing literature. Unfortunately, most student writers are not prepared by their technical courses to conduct such searches. This lack of preparation was recently demonstrated on our own campus when a survey of 167 seniors in engineering revealed that only about eight percent had used or even looked at the basic research sources in engineering. Technical writing teachers, therefore, find it necessary to teach their students about technical information resources. Admittedly, the task is difficult, especially for traditionally prepared instructors who generally have little background in technical subjects or their research sources.

Thus, the purpose of this paper is to provide an orientation to such instruction and to encourage technical writing teachers to seek additional help from the technical librarians at their institutions. We will describe two approaches to providing instruction in the use of technical information resources, approaches that were developed through the cooperation of technical librarians and technical writing teachers at the University of Michigan. Finally, we will suggest adaptations of these approaches for introductory level classes and for schools where information resources are relatively limited.

The First Approach: Library Overview

In the first approach, an overview of relevant library resources is given to all students in a senior-level technical writing class just before they are to write their first technical reports. This overview deals with the following topics:

1. The use of the card catalog, especially the information given on catalog cards, and the use of the shelflist and serials records

2. A description of major indexing and abstracting tools, including *Applied Science and Technology Index*, *Engineering Index*, and *Chemical Abstracts*

3. An overview of government document and technical report literature and the use of *Government Reports Announcements and Index*

4. An introduction to patents literature, major business and technology reference works such as handbooks and encyclopedias, and U.S. standards and specifications, including those set by OSHA and Underwriter's Laboratories

In the technical writing class itself, students are asked to use and cite library resources where relevant. For instance, when they prepare job letters and resumes, they are asked to use several major business resources to find information about their prospective employers: for example, Standard and Poor's *Register of Corporations*—to learn about major product lines, officers, locations, and subsidiaries—and Funk and Scott's *Index*—to locate news articles about the company from such sources as the *New York Times*, *Business Week*, and the *Wall Street Journal*. In addition, annual reports are used to evaluate a company's corporate assets. At other times, students are asked to find relevant material for reports by using several of the sources described in the overview lecture.

Because some students have had little previous library experience, these orientation lectures should be held in the library so that the location of the materials discussed can be indicated. For those instructors who prefer to hold library instruction in their own classrooms, a videotape with handout is available. It is also important to distribute a detailed handout of examples from the sources discussed, a handout that students can use as a reference throughout the semester.[1]

Student response to this approach has been very favorable.

Although many students were overwhelmed by the volume of material with which they were unfamiliar, they were nevertheless grateful to learn that such materials existed. A few students who had previously been assigned technical research papers but had been given no instruction in using library resources regretted that this overview had not taken place earlier.

The Second Approach: Minicourse in Technical Resources

The second approach is based on a one-credit course in the use of library resources. To ensure maximum benefit from the course, students are allowed to enroll only if they have a current research topic for a technical course, for thesis work, or for a project arising out of part-time employment. The course meets for one two-hour session a week for eight weeks and covers the following areas.

1. Books: a discussion of publishers and publishing patterns in technology, the introduction of bibliographic tools such as *Books in Print* and *Forthcoming Books*, and an explanation of the card catalog.

2. Journals: a discussion of university presses and scholarly and popular publishers and the introduction of *Ulrich's International List of Serials* and the national union lists.

3. Indexes and Abstracts: an introduction to the organization and use of subject indexes (e.g., *Applied Science and Technology Index*) and multiapproach abstracting services listing by subject, author, and sometimes by chemical formula, patent number, and author's corporate or institutional affiliation (e.g., *Chemical Abstracts, Computer and Control Abstracts*).

4. Conference Proceedings: a review of the indexing and bibliographic control of conferences through such sources as *InterDok: Directory of Published Proceedings, Proceedings in Print*, and professional associations.

5. Documents and Technical Reports: a review of government, university, and corporate publishing and the distribution of their reports through the office of the Superintendent of Documents and its *Monthly Catalog*, through the National Technical Information Service (NTIS) and its *Government Reports Announcements and Index*, and through private mailings.

6. Standards, Specifications, and Patents: a presentation of the history and development of voluntary and mandatory standards and specifications from American National Standards Institute (ANSI), professional societies, International Standardization Organization, and the United States government, including procedures for obtaining a patent and the use of the literature through searching.

7. Reference Sources outside the Library: a discussion of directories for locating experts (e.g., *Encyclopedia of Associations, Directory of Industrial Research Laboratories*) and the use of yellow pages, city and state offices, direct mail advertising, etc.

8. Business Sources and Updating Skills: the introduction of *Business Periodicals Index*, Standard and Poor's *Register of Corporations, Thomas' Register of American Manufacturers*, and Funk and Scott's *Index;* a discussion of techniques for keeping current and learning about new fields through general scientific periodicals such as *Science, Scientific American*, and *American Scientist;* a review of the publications from various professional societies (IEEE, SAE); and an orientation to the use of industrial libraries.

9. Future Information-Handling Technologies: a videotape presentation on the advantages and limitations of computerized bibliographic searching, including on-line computerized bibliographic search through the Lockheed *Dialogue* or SDC *Orbit* systems for each student. The data bases include on-line files of approximately fifty printed sources, including *Chemical Abstracts, Engineering Index*, and technical report literature from the National Technical Information Service.

Handouts of illustrative materials and bibliographies have been prepared for the various syllabus topics, and one and one-half hours of each two-hour session are usually given over to lecture. Thirty minutes is scheduled as a supervised lab session during which students use the materials introduced in the lecture and the instructor answers questions and offers suggestions. In addition to lecture and lab sessions, the course includes a term project that is explained to students as follows:

1. Select and define a research topic. Examples from past terms include "function and reliability of the emergency core cooling system in a pressurized water reactor," "continuous

casting of high alloy metals, concentrating on available technology and patent applications," and "construction of concrete dams using present methods and techniques."

2. Develop a bibliography using sources discussed in class, including the card catalog (with a list of pertinent subject headings), handbooks, manuals, indexes and abstracts, and patents and standards. In addition, identify a professional society or association whose members would have research interests similar to yours and list conferences this society or other groups have held that are related to your topic.

3. Include with your bibliography comment about the specific value of the materials consulted, difficulties you encountered in searching, etc.

The format and content of the reports vary according to the project. For example, research on pile pullout, a civil engineering problem created by ice and fluctuating water levels, yielded a small number of highly specific citations from the U.S. Army Cold Regions Laboratory and Canadian document sources, the American Society of Civil Engineers indexes, *Engineering Index*, and conferences published by the International Association of Hydraulic Research. In contrast, research on microcomputers and microprocessors provided pages of sources, and the project developed as an evaluative essay on major tools for further research. The introductory portion of one student's report is given below.

> I researched the subject "Cost of Concrete Dam Construction" at the University of Michigan. My objective was to compare the cost of different construction methods for various types of concrete dams. Because of my limited knowledge of the subject, I needed to determine the types of concrete dams and the methods of construction before attempting to determine costs.
>
> I consulted the following sources to find relevant references on which to base my term paper:
> 1. Card Catalog (Engineering Library)
> 2. *Engineering Index* (Engineering Library)
> 3. Card Catalog (Graduate Library)
> 4. *Comprehensive Dissertation Abstracts* (Graduate Library)
> 5. *Applied Science and Technology Index* (Engineering Library)
> 6. *Government Reports Announcements and Index* (Engineering Library)
> 7. American Society of Civil Engineers *Publications Index*
> 8. Browsing

The report continued with a description of methodology and an extensive bibliography that ranked the relevancy of the materials found and described the values and limitations of the various sources consulted.

Evaluation is an important part of the course because it encourages comment on the technical content of project materials and tends to reinforce the merit of the course itself. For example, in evaluating the above report, Professor Robert Harris, Department of Civil Engineering, noted: "I found [this student's] report very relevant to the subject chosen. . . .The breadth of sources cited was good and his evaluation of their potential for his study was carefully made." And the student responded: "I feel Engineering Humanities 420 was very valuable and saved me a great deal of time and trouble. I would highly recommend the course to anyone who has trouble in research or the use of the library in general." Typically, faculty and student evaluations have been favorable. Some have said, "I think this course would be most valuable to every student," "I found the course very stimulating and interesting," and "I have learned more things that I will use and have enjoyed this course more than any class I've ever taken."

Adapting the Two Approaches to Other Settings

Instructors whose library resources are limited can, of course, follow the lecture or course syllabus but introduce fewer resources. They might also look to resources outside their campuses. For instance, teachers in larger cities can rely on public libraries to provide most of the materials needed since these libraries usually serve small business and industry in the area and have appropriate materials available. Some teachers may have access to nearby colleges or universities for supplementary source materials. Others may need to rely more heavily on handouts and borrowed materials, including audiovisual aids and materials obtained through national interlibrary loan or State Access Office programs. Instructors in two-year colleges or those teaching at the freshman or sophomore level might choose to concentrate on less technical sources such as *New York Times Index, Consumer Index, Applied Science and Technology Index* and city, county, and other government agencies near their institutions. All teachers should note that it is important to teach a few tools thoroughly, stressing general concepts and the transferability of research skills. The student who learns how to use two or three indexes and abstracts effectively will quickly learn the use of others when he or she has the need and the appropriate materials available.

Teachers of technical writing may find several rewards in teaching—or enlisting the librarian to teach—the use of technical resources. They increase their interaction with technical professors (whose students they share) as well as their own knowledge of and respect for various technical fields. They may also earn more respect from their students by demonstrating an unexpected degree of competence in the technical fields of their students. Finally, technical writing instructors who teach research skills contribute significantly to the professional development of their students.

Notes

1. This handout as well as other handouts and videotapes referred to in this article are available from Maurita P. Holland, Head, Technology Libraries, Engineering-Transportation Library, The University of Michigan, Ann Arbor, Michigan 48109. More detailed information about the resources cited in this paper and about additional resources is available in Robert H. Malinowsky et al., *Science and Engineering Literature: A Guide to Current Reference Sources*, 2d ed. (Littleton, Colo.: Libraries Unlimited, 1976) and in K. W. Mildren, ed., *Use of Engineering Literature* (London and Boston: Butterworths, 1976).

12 Technical Illustration

Charles E. Beck and William J. Wallisch, Jr.
United States Air Force Academy

In some technical writing courses, instruction in the design
and preparation of visual aids receives only indirect attention.
Frequently the topic is subsumed entirely into other types of
instruction or is taught only theoretically, almost as an after-
thought. A lecture and workshop model for teaching technical
illustration is outlined here.

Most technical documents require visual support, and technical
writing courses must, therefore, include substantial instruction
on how to design and use visuals. In this paper we describe a
lecture-workshop model that can be expanded over several class
periods. Indeed, we recommend that the single lecture and workshop described below serve as a starting point, a minimum requirement, for more systematic instruction in the design and
preparation of visual materials.

Background

Students have been subjected to visual presentations all of their
lives. A potent early form probably came by way of television
advertising, with magazine and newspaper advertisements following.
As the products of well-backed efforts to persuade audiences,
these media presentations almost always rely on the effective use
of graphics, design, color, and form. But students in technical
programs may be even more aware of visuals than their counterparts in nontechnical programs. Engineering and science majors,
for example, use technical texts that contain hundreds of graphs,
charts, and visuals of all descriptions. Technical students are
caught up in, preoccupied by if you will, a scientific environment
that provides them with graphics at every turn. Our point is that
technical students may already have a very good idea about what

makes a good visual: the world of advertising has subjected them to many examples, and their scientific courses have offered effective visuals dealing with theory and practice.

All of this exposure provides easy entry into an intitial assignment, and you might introduce the visuals workshop by asking students to gather both good and poor examples of visual presentations from books, magazines, and journals. Such an exercise allows students to take part in the opening discussion and, at the same time, clarifies and refines their visual standards.

The Lecture

You might begin your lecture by discussing the impact of visuals: they break up the text, providing relief from the verbal presentation; they provide variety, adding interest to reports; they reinforce subject matter, increasing retention. You might also note that the care with which a technical writer has handled visuals disposes the reader favorably toward the report, perhaps because the reader assumes that the care used in preparing the visuals was also used to write the report. That assumption may be false, but unconsciously most readers make it. No matter how you introduce the topic, certain basic concepts should be included in the lecture:

1. *General.* Visuals reinforce but do not replace the text. Consideration of the audience applies to the selection of visuals just as it does to the selection of language for the text. Since placement on the page has a major impact on the audience, students should carefully analyze layout.

2. *Photographs.* Poorly planned and executed photos clutter and confuse. Students, therefore, should prepare simple photos, label them with essential information, and mount them securely to ensure a professional appearance in the body of the text.

3. *Drawings.* Students should label drawings according to perspective; unlabeled views cause trouble for readers.

4. *Diagrams.* Since simple diagrams provide an effective way to reinforce ideas, they should be used liberally. However, students should be cautioned about the use of complex schematic diagrams that are too complicated for most audiences to follow. Simple schematics belong in the body of the report; complex, exact diagrams belong in an appendix.

5. *Tables.* Since tables actually provide rather than merely reinforce data, they should be treated separately from other visuals. For example, tables are labeled at the top, since the reader considers the table as part of the basic text. But even tables are not totally self-contained, and students must provide an adequate discussion of tables in the text so that the reader can understand and interpret the material.

6. *Graphs.* Detailed graphs on lined paper should be placed in the appendix; in the body of a report, a more general indication of relationship or correlation will do. In all graphs, vertical scales should be placed on the left side of the graph, horizontal scales, across the bottom of the graph; each scale should begin from the bottom left-hand corner. If the graph contains multiple lines, the most important line should be placed in bold print, the second most important line in light print, the third in dashes, and the fourth in dots. Since graphs show relationships between known standards and new information, they are particularly useful for an uninformed audience. The greatest danger with graphs, however, is the selection of a misleading scale, and students must ensure that their portrayal of data is ethical.

This lecture should be accompanied by its own visual reinforcement, of course. As you discuss each type of visual, use slides and transparencies that illustrate effective and ineffective uses, the more up-to-date the better. With lead time, most instructors can rely on the graphics division in their own institutions. These professionals can make very attractive copies of visuals in either transparency or slide form. Often, again given the time, they can create original visuals that illustrate your points more effectively than visuals you have collected from other sources.

Certain guidelines for designing effective visuals should also be emphasized in the opening lecture and reinforced with examples:

1. Keep visuals simple: too much detail confuses and distracts.

2. Use sharp colors for contrast. Pencil is too light for reports or briefings, and pastel colors are too weak.

3. Use minimal wording and short labels: too many words counteract the effect gained by visual reinforcement.

4. Differentiate labeling arrows from process arrows by making those for labels smaller.

5. Adapt visuals from books or magazines, but be sure that they are appropriate for the intended audience.

6. Use good quality materials.

7. Never interrupt a sentence with a visual; a visual and the text should be an integrated, logical unit.

8. If a visual in a typed manuscript requires an entire page, place it on the back of the page that precedes the discussion of that visual. In this way, the visual directly reinforces the text.

9. Do not try to get by with minimal effort. Well-prepared visuals enhance the readability of reports and improve communication.

The Workshop

There is obviously more material on the preparation and use of visuals than can be covered in a single lecture. Most instructors have collected a wide range of visuals and some will want to discuss special cases, for example, briefing visuals. (Since briefing is part of the course requirements at the Academy, our students need to see examples of large, effective briefing aids.) You might, therefore, consider additional lectures before you turn to the visuals workshop.

When you do move on to the workshop, spend the bulk of the time in the workshop mode; that is, give students the opportunity to make and critique visuals. Soon they will be designing them for assignments and receiving your criticism of their efforts. The workshop gives them practice and allows for "free" mistakes. Along with your suggestions, they will also obtain valuable peer feedback on their work.

We ask students to bring the following materials to the workshop: ten sheets of unlined paper, three sheets of graph paper, a variety of colored pens and pencils, a small ruler, and a geometric compass. We provide a set of statements dealing with statistics, descriptions, and mechanisms that can be put into visual form. These can be excerpted from student papers, textbooks, and professional reports.

The sample set that follows was taken from student reports. These excerpts are obviously in need of visual clarification, and the student's task is to produce easily understood visuals that enhance the reader's comprehension. It is important that you have effective visual solutions in mind, and typical instructor reactions follow each excerpt. You might also want to have on hand the visual that originally accompanied the statement as you found it in a text or report.

Directions to Students: In considering illustrations for a text, remember that visuals *reinforce* the author's text; they do not *replace* it. The question is not "When must I use a visual?" Rather, the question is "When will a visual help my reader understand my report?" Therefore, writers must examine their texts for places where visuals clarify ideas.

Below are excerpts from five student reports. As your instructor directs, read an excerpt and design an appropriate visual on a separate sheet of paper. These visuals may take many forms: flow charts, schematic diagrams, tables, line graphs, pie charts. Use your imagination to devise new techniques. Although we do not expect you to be a proficient artist, you should be able to provide appropriate and effective visuals for the excerpts given here. Since all of your papers after the description of a mechanism include visuals, we hope these exercises will improve your ability to create ones that are effective.

Excerpt 1: Wind Tunnel Experimental Data

This table, taken from a larger table found in the appendix of a report, contains exact data points. Using the data, construct two graphs. The first should reflect the data as it appears in the appendix; the second should reflect the data as it would appear in the body of a report for a general audience.

Results of Bomb Data

Mach Number	Coefficient of Drag
0.14	0.0672
0.25	0.0682
0.37	0.0682
0.48	0.0626
0.61	0.0727
0.71	0.0756
0.82	0.0848
0.86	0.0856
0.91	0.0932
0.97	0.0997
1.04	0.2710
1.13	0.3031
1.20	0.3029
1.44	0.3460

Instructors should anticipate the following kinds of problems as students handle excerpt one.

General. Extensive data given with such precision usually appears in the appendix of a report. But the appendix may also contain a detailed graph of the data. This assignment asks students to provide a graph for the appendix as well as a more general version for the body of the text, where the reader needs to discern relationships but does not need precise detail. Some students find it difficult to determine proper spacing on the graph paper.

Minimum requirements. Two graphs, one on graph paper with precise spacing of data, the other a more general graph showing the relative position of the curve between data points.

Alternate possibilities. Some students might use a bar graph rather than a line graph; math wizards can find an equation to fit the data curve (if they bring calculators), but such an approach adds little for the general audience.

> Excerpt 2: What People Dream About
>
> Calvin S. Hall studied 10,000 dreams to find out what people dream about. Then he classified the dreams according to the following categories: dream setting, cast of characters, plot, dreamer's emotions, and color. A summary of his findings follows: parts of a building comprise 24% of the dream settings, automobiles 13%, entire buildings 11%, and recreation areas 10%. The cast of characters involves strangers in 43% of the dreams, friends in 37%, and family in 19%. The plot involves movement in 34% of the dreams, talking in 11%, sitting in 7%, and socializing in 6%. The dreamer's emotions include apprehension in 40% of the cases, anger, happiness, or excitement each in 18%, and sadness in 6%. Finally, color applied to only 29% of the dreams.

Instructors would do well to expect the following problems as students tackle excerpt two.

General. This example shows the problem created when a writer incorporates a number of statistics into the text. The reader becomes saturated with numbers and misses the significance of the data. Further, as the data is given here, the categories do not always add up to one hundred percent because the author classified only those responses that fit selected patterns, ignoring less typical responses. Most students should have little trouble developing a simple table for the data, but they will have trouble developing additional visuals; require students, therefore, to do more than present the data in a table.

Minimum requirements. An extended table listing the data within appropriate categories and a visual in which the distinctions between categories are readily apparent.

Alternate possibilities. The data could be presented in a series of bar graphs; the bars would provide a means of comparing across categories. Breaking the information further, students can draw a series of pie charts, one for each category; highly imaginative students can add a picture for each breakout within a category, but the sketch would be a mere outline if done in a class period.

Excerpt 3: Electrochemical Aspects of Corrosion

All forms of corrosion have in common an electrochemical reaction requiring four elements: an anode, a cathode, an electron path, and an electrolyte. The anode describes the metal or part of the metal where the actual corrosion or destruction of material takes place. A process called "reduction reaction" takes place at the cathode. The electron path is a metal or alloy that conducts electrons from the anode to the cathode, and the electrolyte is the environment that contains conducting anions (negative ions) and cations (positive ions).

The reaction occurring at the anode, or the anodic reaction, given metal "A" is generally written

$$A \rightarrow A^{+z} + ze^-$$

In other words, atoms at the anode give up electrons. These electrons travel through the electron path to the cathode and are used up in the reduction reaction.

The reduction reaction occurring at the anode is characterized by the equation

$$ze^- + M^{+z} \rightarrow M$$

Thus electrons from the anode travel to the cathode through the electron path and combine with an ion in the electrolyte (M^{+z}) and leave the metal.

On a metal surface anode and cathode sites exist randomly and may change instantaneously. In other words, anode and cathode sites change rapidly so that the above process occurs all over a metal surface.

Instructors should be alert to problems like these as students deal with excerpt three.

General. This text provides a basic discussion of electrical potential and the flow of electrons. For an electrical engineering major, the text is simple, but for the nonspecialist, the text is more intelligible with visuals. Be prepared to argue with electrical engineering majors on this one. Nonmajors may have a difficult time with this text, but let them work with it a bit. The text reads fairly well and presents the concepts in a simple form. With a little effort, even the novice can understand and work with this text.

Minimum requirements. A simple schematic depicting the main parts (anode, cathode, and electrolyte) and the direction of movement. Terminology must match the discussion and equations in the text.

Alternate possibilities. An imaginative cartoon depicting ions traveling from one point to another within a solution.

Excerpt 4: Solid Fuel Launch Vehicle Propulsion

Thrust is basically a force that causes an object to speed up or
slow down in some direction. By producing thrust, a launch
vehicle propels itself and its payload upward against the force
of gravity. When fuel burns, hot expanding gases cause mass to
travel at high velocities, producing a tremendous force that acts
in a direction away from the combustion. By Newton's Second
Law, which states that for every action there is an equal and
opposite reaction, the thrust force acts in a direction opposite
that of the escaping gas.

In order to move away from the earth's surface, a launch
vehicle must produce sufficient thrust to overcome gravity.
Gravity acts in a direction toward the earth's center and causes
objects to remain on, or fall toward, the earth's surface. To
achieve takeoff, the vehicle engine produces thrust greater than
the opposing gravity force.

Steering a launch vehicle as it travels upward against gravity
involves changing the thrust direction. Imagine a line drawn
vertically through the center of the cylindrical launch vehicle
or rocket. Altering thrust direction to the left or right of the
center line causes rotation around a point in the vehicle called
the center of mass. A thrust direction right of the center line
causes counterclockwise rotation, while thrust direction left of
the center line causes clockwise rotation.

Instructors might anticipate the following problems as students
work with excerpt four.

General. Expect complaints that this text needs no visual because the concepts are obvious. For technical students, the text is
obvious, but not all audiences have a scientific background. Stress
visuals as reinforcing the text, not as replacing or carrying it.

Minimum requirements. The third paragraph needs a simple
schematic indicating the direction of motion, the center of mass,
and the direction of rotation.

Alternate possibilities. Rather than relying on a simple diagram,
students can graphically develop the third paragraph with model
rockets, drawn to scale. Also, the earlier paragraphs could include
simple lines and arrows to depict relative forces for thrust, gravity,
and so on.

Excerpt 5: Pollutants

Pollution from automobiles exists in basically three forms:
hydrocarbons, carbon monoxide, and oxides of nitrogen.

Hydrocarbons, designated as HC, are substances composed of
carbon and hydrogen. Gasoline used in most cars, known as
octane, is chemically named 2, 2, 4-trimethypentane. The molecule of gasoline exists as a chain of five carbon atoms bonded
(attached) to each other with three other carbon atoms bonded
to the second and fourth carbon atoms in the chain. Because the

carbon atom has the capability of bonding in four locations, hydrogen atoms bond around the eight carbon atoms so that each carbon atom has four bonds or four groups attached to it. If combustion were complete, this molecule would react completely with oxygen to produce water and carbon dioxide. However, complete combustion does not occur within the engine, so the gasoline molecule breaks up into different substances of carbon and hydrogen known as hydrocarbons. . . .

The last pollutants are oxides of nitrogen designated as NO_x. The automobile engine produces NO, NO_2, and N_2O_4, where the N stands for a single nitrogen atom, the O stands for a single oxygen atom, and the number represents the number of atoms of nitrogen or oxygen. The amount of NO_x produced does not depend on incomplete combustion but on the temperature of the combustion. In an engine, the quantity of NO_x produced increases significantly at about $4000°F$.

The fifth excerpt may well raise the following issues during the workshop.

General. The text presumes an elementary knowledge of chemistry. For classroom purposes, the paragraph on carbon monoxide was omitted. The text is technical but should cause no real problems. Students may have to dust off their knowledge of chemistry, but the text is clear enough and any student should be able to develop visuals for the concepts.

Minimum requirements. Two visuals: the first using letters and lines to depict the chemical bonds (H–C–C, etc.) and the second based on a line graph showing the relationship between temperature and oxides of nitrogen. Although the last paragraph is vague concerning the precise movement of the line, students should be able to show a general relationship.

Alternate possibilities. Students with better backgrounds in chemistry can depict the variety of hydrocarbons formed through combustion. A picture of the auto exhaust with various pollutants emitted into the atmosphere might enhance the discussion.

Initially, then, students may feel uncomfortable during the workshop and puzzled by how to go about developing visuals that support the text of the five excerpts and help the reader. You might ask each student to draft a visual for one of the excerpts and then divide the class into groups of three or four to compare efforts. This interchange helps students consider alternate approaches to the same design problem. After students have shared ideas, let them work out one or two visuals of their choosing. To demonstrate your own involvement in the workshop, you might try your hand at turning out a visual or two. In any case, be prepared to monitor the performance of each student, offering comments and suggestions.

Although students may work alone or in groups, questions and chatter are encouraged. Students should be asking for your help and nudging classmates for advice and feedback. The workshop session should be lively; there should be noise, an exchange of ideas, and some good finished products.

Expanding the Lecture-Workshop Model

Obviously we have given you only the bare bones of a lecture-workshop in the design and preparation of visual materials. Examples gathered by students might form the basis of a third session. Another session might be spent discussing transparencies or slides made from the workshop efforts of students. Or perhaps you could invite a technical artist from your institution or from industry to speak, or include these professionals in other ways in your workshop. The author of a recent scientific text could provide valuable insight into the problem of constructing clear visuals, and the editor of a technical journal or industrial publication would be an extremely interesting speaker. We should look to industry for speakers, examples, and general reinforcement of our curriculum. Industry is, after all, the destination of most students after training.

And the industrial tie-in is appropriate in other ways, even when speakers acknowledge that they do not actually prepare visual material themselves. Like the dentist who must learn to make dentures in school but orders them from a specialty laboratory after graduation, so must our students develop a knowledge of what they want in visuals even if they do not design their own visuals later on. They must, however, know what they want before they can order it, and often they must sketch ideas for the commercial artist who will do the work. Hopefully, your lectures and workshop will arm your students with that knowledge and the ability to express it to someone else.

13 Engineering Students Write Books for Children

Herman A. Estrin
New Jersey Institute of Technology

The author proposes an unusual project for students in technical writing courses: authoring short, science-related books for children. Meeting the special requirements of young readers emphasizes the need to consider audience whatever the writing task.

"I like *Ouch, Gravity Hurts* because it tells how gravity works, and especially I like the part where the apple fell on the boy's head." "*Garbage: Here Today and Gone Tomorrow* was an exciting book to read because the pictures and words were neat and understandable." Comments like these were made by fourth-graders at Plainfield, New Jersey, elementary schools after reading science books written by civil engineering students of New Jersey Institute of Technology. Nearly forty science books authored by future engineers were given to the Plainfield Public Library to be read by young readers. The engineering students were enrolled in a technical writing course that emphasized how writers adapt their writing to an audience. Civil engineers must write for different audiences—supervisors, other engineers, technicians, and the public. The instructor decided to stress the importance of intended audience by focusing on children, and the student-authors responded by using graphics, humor, figures of speech, sensory details, and vocabulary suitable for youthful readers.

Initiating the Project

The instructor began by presenting the engineering class with ten books from the Thomas Y. Crowell series "Let's Read and Find Out": *Before You Were a Baby, High Sounds, Low Sounds,*

Your Skin and Mine, Why Frogs Are Wet, At the Drop of Blood, Shrimps, Hear Your Heart, How You Talk, Ladybug, and *A Book of Mars for You.* In addition, students were encouraged to answer these questions as they read the books:

Content. What is the overall content of the book? How would you evaluate the scientific information in the book? Is the subject matter timely, relevant, and useful to the reader? Comment on the introductory matter and the conclusion. Does this book reflect needs and experience common to all of us and yet have spontaneity and freshness?

Format. What are the dimensions of this book? How many pages does it contain? Describe its print. Does each page have "white space," or is it filled with printed matter? How is color used? Discuss illustrations, pictures, charts, and graphs. Can you recommend additional graphic aids? If so, what kinds, and why would you recommend them?

Style. What kinds of sentences are used throughout the book? Does the author use figures of speech, parallel structure, judicious and effective repetition? Cite examples. How would you evaluate the vocabulary of the book? How does the author explain difficult terms? Does the author make the reader use both eyes and ears to understand the context of the book? How? Children love the tastes, the smells, the colors of things. Does the author use as much sensory detail as possible? How would you use more sensory detail? Does this book have humor? Is it direct and obvious?

Evaluation. If you were to write a book on this subject, what approach would you use to arouse the interest of the reader? How could this book be improved? Did you learn any new information from the text? Did the author reach the audience level for which he or she was writing?

Selecting a Subject

After this review of children's science books, students were asked to select a subject about which they had a thorough knowledge and which they thought would capture a child's interest. Mel Cebulash of *Scholastic Magazine,* a guest speaker to the technical writing class, advised that children are especially interested in subjects concerning ecology, astronomy, and geology. Also, they like books about animals—snakes, frogs, bees, shrimp. In addition, children are interested in books about the construction trades, science mysteries and riddles, and science fiction. However, students were cautioned to find a subject in which they themselves were especially interested. Then they wrote their books. The titles of these student-authored books for children are found in Figure 1.

Distribution and Reaction

The completed books were then circulated in the children's division of several New Jersey libraries. One library lent them mainly to elementary school teachers who used them in their classrooms. One teacher reported that a student said, "This is the first book I read from cover to cover." A children's librarian wrote, "If a child is interested in something particular, he will

Science

Let's Learn about Electricity
Kinds of Telescopes
Highways
The Children's Book of Roads
Earth's Friendly Blanket
What Makes It Rain?
Rocks Are Everywhere
On a Journey to the Moon
Did You Ever Look
    under Your Street?
Ouch, Gravity Hurts!
Why You Can Build Sand Castles

Ecology

What Is Water Pollution?
Ocean Ecology
Where Is down the Drain?
Our Beautiful Dams
Traffic Jams
The Solution to Pollution
Where Are Our Beaches Going?
A Loud, Loud World:
    Noise Pollution
Alice in Garbageland
Garbage: Here Today and
    Gone Tomorrow

Hobbies

The Soccer Game
On the Art of Fencing
How Would You Like to Travel?
Skin Scuba Diving
How to Keep Tropical Fish
Johnny Learns How to Take
    a Picture
Charlie Goes Skiing
The Guitar: Lessons in Tuning and
    Playing the Basic Chords

Guidance

The Surveyor
The World of Civil Engineering
Eddie Electron
I Am an Electron
The Engineering Tree
Big Bridges, Little Bridges
Ants: The Tiniest Engineers
The Adventures of an Engineer
What Does a Civil Engineer Do?

Figure 1. Titles of student-authored books for children.

pursue that—no matter how many copies of *Charlotte's Web* we have here. But beyond the basic interest, it is a book's presentation which attracts a child."

Comments by engineering students mentioned the challenges: "In writing a child's manuscript, I found it very difficult to adapt the words to a child's level of reading. . . . It was a challenge to communicate this information on a lower reading level." Another wrote, "I found that reviewing published books on science literature for children gave me a better insight into what is expected of a writer."

Some engineering students, believing that children had few ideas about the various branches of engineering and about the duties and responsibilities of an engineer, produced guidance books about the profession aimed at arousing interest in becoming engineers. Since the Engineers' Council for Professional Development is interested in guidance information, several of these books were submitted to that association. Executive Director David Reyes-Guerra agreed to print them because they were specific, concise, attractive, and inspirational. He felt that the engineering students had proved to be up-to-date about the profession and that the books were not only informative but sincere and enthusiastic as well. Business and industry might also find the books useful in communicating their services to young people and in shaping the public image of engineers.

In short, this assignment has proved a useful one in two senses: it produces a product that other people find valuable, and it helps students define the needs of a special audience and develop skills to meet those needs.

14 Teaching the Writing of Instructions

Donald H. Cunningham
Morehead State University

John H. Mitchell
University of Massachusetts

The writing of instructions should be included in a technical writing course regardless of the level of the course or the makeup of the class. It allows students to write without extensive preparation, to master a format related to on-job assignments, to write to an identifiable audience with the specificity and clarity necessary to all technical writing, and to use graphics. A strategy for introducing students to the writing of instructions is discussed and three specific assignments are provided.

Vocational, technical, and science students for the most part are bright, eager to learn, willing to work, and performance-oriented. They grow restless when they are merely "told" about writing, and they certainly do not want to hold theory in their heads for a long time before they see it in practice or practice it themselves. Thus, the presentation should be "hands on," and students should be allowed to arrive inductively at the following general principles:

1. A reader receives more information and understands it better when the information is in an expected and familiar form.

2. A reader must get only one meaning—the correct meaning—from a statement.

3. The writer must identify the audience so he or she can analyze the "teaching content" of the writing at hand.

4. A message loses part of its meaning in transfer.

An effective way to begin is to ask students to discuss instructions they have used with kit assemblies of model cars and airplanes, with sewing patterns, cookbooks, and lab manuals. They will remember, usually quite vividly and precisely, the successes and failures they experienced. They will also be able to identify the level at which certain instructions were written, the sources of problems in following specific instructions, and the devices that helped them most in following a set of instructions.

Another way to open (or to follow up the previous discussion) is to put students in the reader's shoes by asking students to follow a set of instructions in the classroom. Writers need to be reminded what it is like to be on the receiving end of the instructions. Bring in two examples of instructions—one bad, one good. The poorly written instructions allow students to experience how readers feel when they anticipate being told how to do something and are then let down by poor writing. The moral of the well-written instructions is clear.

We usually turn to the badly written instructions first, asking three volunteers unfamiliar with the task to turn their chairs to face the class so their classmates can observe them as they try to follow the instructions. The three should also position themselves so they cannot see one another. Everyone is given a copy of the instructions, but only the three volunteers try to follow the procedure. We station ourselves nearby, for we have the materials the three will need in a big box. Below is a set of instructions for making a paper rose that is guaranteed to cause problems quickly.

Materials: bathroom tissue, paper clip, pipe cleaner

Directions:
1. Take ten sheets of tissue paper.
2. They should be placed directly on top of one another.
3. Fold them in a fanlike manner by just turning up and back.
4. A paper clip must be placed in the middle of the fold.
5. Separate each tissue by taking the top one, pulling it toward you. Do the same to all ten sheets.
6. The paper clip should be removed and the pipe cleaner placed in its place to form the stem.

The general terms in the materials list create difficulties immediately. The secret here is to have in the big box as many different kinds of paper clips, pipe cleaners, and tissue as possible. It is easy to produce a maddening array of paperclips of different sizes, materials, and configurations. Pipecleaners can range from blades on pipetools that smokers carry to small bottles of chemicals. A poll of the class will reveal confusion about what "bathroom tissue" is. Once you establish that toilet paper is what should be specified, you can produce rolls and folded bundles of it.

As the three students read on, they are likely to run into further difficulties caused by the conditional statements and the inappropriate use of passive voice. We think the problems and their solutions are fairly obvious, but we do want to make two additional comments about using these poorly written instructions.

First, students delight in providing further examples of ambiguity in the simplest of statements. One told of a sign outside a campus dining room that read "Now Serving Faculty and Staff." Another remembered seeing a sign that stated "Shirts and Shoes Required to Eat in Dining Room." Someone had added the note, "Trousers and Socks Can Eat Where They Please." A colleague once told us that her instruction to "Write on one side of the page only" was misunderstood by a student who drew a line down the middle of the page and wrote to one side of it. The student assumed that the other side was for the instructor's comments. All of this, zany as it might get, is instructive. The last example permits us to comment on instructions so brief that readers are permitted to think for themselves. You may have to put a lid on the discussion, but students will have seen how slippery language can be, and you will have proved Murphy's law—if something can be misread, it will be misread.

Second, sometimes a student transcends the poorly written instructions and creates a tissue rose, but not without the rest of the class seeing the frustration and trial and error. The point can be made that for every reader who can figure things out, dozens will not. Another point can be made by reminding students that some high school graduates, according to governmental studies, are so inept at reading that they cannot follow instructions on a road map. If you wish, you can introduce the role of graphics here.

This exercise leads students to recognize the cause of most unclear instructions: writers who write about a procedure they know are likely to explain just enough to remind *themselves* of the specifics of the procedure. They are shocked to learn that what they write doesn't always communicate to everybody else.

A competent set of instructions is given below. Introduce it with a few generalities on legal aspects that will startle students: Writers are legally responsible for their own words. Should technicians injure themselves, damage their equipment, or destroy material while following a set of published instructions, they have the option to sue the writer or the publishing agency. The technician will win the suit if he or she can show that the instructions failed to caution a technician with no greater skill than the ability to read.

HOW TO CLEAN A 20-GAUGE MODEL-69 SAVAGE
OVER-AND-UNDER SHOTGUN

PROCESS:

Essentially, cleaning a shotgun involves dissolving nitrate from the barrel, brushing all foreign matter from the barrel, and coating all metal parts of the gun with a thin film of light machine oil. A variety of commercial solvents is available, and either Hoppe's Number 9 or J.C. Higgins Nitro-Solvent is recommended.

TOOLS:

    a. One 20-gauge, 3-piece cleaning rod
    b. One bronze cleaning brush for cleaning rod
    c. One slotted tip for cleaning rod

MATERIALS:

    a. One 8-ounce bottle J.C. Higgins Nitro-Solvent
    b. One 8-ounce bottle J.C. Higgins Siliconized Gun Oil
    c. One box J.C. Higgins cleaning patches
    d. One medicine dropper
    e. One silicon wiping cloth, 12 inches square

GENERAL WARNING:

Always assume a firearm to be loaded until you have checked it yourself; never point it at anyone, and never work on a loaded gun.

DIRECTIONS:

    1. Point gun down with muzzle 6 inches from floor.
    2. Push safety up until "S" is uncovered.
    3. Push breaking lever to right and open gun.
    4. Remove shells from barrel chambers.

WARNING: DO NOT WORK ON A LOADED GUN.

    5. Assemble 20-guage, 3-piece cleaning rod.
    6. Screw slotted tip into cleaning rod.
    7. Place fresh cleaning patch in slotted tip of cleaning rod.
    8. Drop seven drops of Nitro-Solvent on cleaning patch with medicine dropper.
    9. Push cleaning rod through lower barrel from breech end and swab entire lower barrel with in-and-out motion of rod.
  10. Repeat steps 7, 8, and 9 until patch remains clean.

    NOTE: Heavy lead deposits in barrel cannot be removed by swabbing with Nitro-Solvent. Should they exist, attach bronze brush to cleaning rod and slide rod in and out until deposit is removed.

11. Swab upper barrel by repeating steps 7, 8, and 9.
12. Place fresh cleaning patch in slotted tip of cleaning rod.
13. Drop seven drops of Siliconized Gun Oil on cleaning patch with medicine dropper.
14. Push cleaning rod through lower barrel from breech end and coat entire lower barrel with light film of oil.
15. Coat upper barrel by repeating steps 12, 13, and 14.
16. Close gun.
17. Drop seven drops of Siliconized Gun Oil on a fresh cleaning patch with medicine dropper.
18. Rub all metal parts of gun with oiled patch until a light film of oil has been deposited on exposed surfaces.

CAUTION: Do not rub oil on wooden stock or forearm. An unsightly darkening of the wood will result.

19. Wipe all fingerprints from surface of gun with silicon wiping cloth.
20. Store gun in dry place.

We have found it effective to discuss this set of instructions first for layout, then for arrangement within sections, and finally step-by-step.

Layout can be introduced by asking if anyone has seen this format before. If you have veterans in class, they will say it looks like a military TO—and they'll be right because the layout is that fixed by Mil(M)–005474C. That can lead you to the following points: (1) Twenty million veterans constitute a conditioned audience, many of whom reject data in an unfamiliar form. (2) Production industries further condition readers by using the same form in the instruction/maintenance manuals they issue with appliances and machinery. They are capitalizing upon a conditioned audience and upon the fact that military TO's are written for readers with little training and less motivation. (3) Do-it-yourself magazines use the form because it permits complete explication and, hence, helps to ensure protection from legal reprisal. What you are trying to do here is to convince students not only that a conditioned audience exists but that readers tend to reject material in nonstandard formats.

The sections can be handled separately. The title is designed to be exclusive and to motivate the reader. The writer is not legally responsible if a technician applies the instructions to anything other than what the title describes. The writer motivates the reader by promising something. If the reader wants a clean gun, he or she will accept the commands (imperative verbs) that begin each numbered step. This is a good place to make the point that imperative verbs can offend readers and are seldom used outside of instruction books.

The process section is an overview explaining what, when, and occasionally why. The why part is placed in the process section rather than in the directions proper where it would clutter and distract. Proprietary terms (Hoppe's Number 9, J.C. Higgins Nitro-Solvent) are used. Conflict of interest, however, exists when a privileged document (material printed by any level or branch of government) implies that a private or proprietary product is best.

The tools section is laid out like a checklist. The order in this instance is not random, and each entry is labeled a, b, c, etc. Here is a good opportunity to discuss audience motivation: a bored technician may skip prefatory material and start work with the first Arabic numeral. That is why numbers are reserved for the specific steps. Here is also an opportunity to teach students to anticipate later writing problems. If they are shown how difficult it is to write directions around a limited number of tools, they will learn to include tools for all normal contingencies. You can illustrate this point by showing the number of steps needed to measure four gallons of water if you have one full can holding eight gallons and two empties holding five and three gallons.

The materials section can be used to discuss the need for specificity. A digression upon cookbooks—which are after all anthologies of instruction—is useful. A good many cooks have pondered over a "handful" of flour and a "pinch" of salt. Cookbooks containing these phrases blunder by forcing the doer to think. No directions should permit the luxury of decision; it is fatiguing in itself and holds the possibility of error. A "handful" is a specific part of a cup. A cook will worry unless told what part.

General warnings may be used to reinforce earlier discussions of legal responsibility. Humorous anecdotes of lawsuits against corporations are more effective than horror stories. Students would rather hear about a suit against Westinghouse brought by the parents of a youngster who was electrically shocked while urinating on a Westinghouse generator than about a suit against a school board following a maiming in a shop course.

The directions section is perhaps best discussed prescriptively as a problem of layout. The audience expects each step to have one number and one command (verb). What then about steps 3, 9, and 14? More important, however, is the physical appearance of a checklist. Each step begins with a number, a period, a space, and an imperative verb. Reader problems can be created by placing the verb elsewhere. The only elements permitted to break the column of numbers are warnings, cautions, and notes. A warning refs to personnel; it is used when technicians might hurt themselves and others. A caution refers to equipment or product; it is used when technicians might damage tools or what they are working on. A note refers to contingencies; it is structured as "if this, then this."

After the introductory experiences with poorly executed and competently written instructions, students are ready to try their hands at the writing of instructions. The three assignments described below provide opportunities for students to apply the principles they have arrived at inductively. Only with "hands-on" practice will they add the writing of instructions to their professional repertoire.

1. How to change the battery in an automobile of the student's choice is an assignment that encourages practice in the points discussed above. Evaluating the instructions is most easily done section by section. The title should specify the voltage of the battery and the year and model of the car. The process —if required—should be straight chronology. The tools should include at least two wrenches, pliers, cable puller, wedge-head screwdriver with insulated handle, wire brush, and fender cloth. The material should include at least a specific battery and a given quantity of distilled water. The general warnings should alert readers to problems of corrosive acids, electric shock, flammable gas, and perhaps weight. The directions—at least twenty steps—should be vertically aligned and parallel in structure: what (verb), to what, with what. Warnings should follow steps involving movement of both old and new acid-filled batteries. Cautions should follow steps replacing cables. Notes should contain contingency information: If you have a Volkswagen, the battery is under the rear seat.

2. How to tie a shoe is typical of topological or motion problems: it can best be explained with visuals. Should you assign a words-only solution, responses can be evaluated through the student's use of metaphor (Form a vertical X by crossing the laces.) and through the clearness of left-right terms (Paint the tip of the left lace red. . . . Hold the red tip between the left thumb and forefinger.).

3. Rewriting an unclear set of directions is an effective way to emphasize the need for exact words and single commands. The passage below is taken from an early do-it-yourself magazine and can be used as the basis for a revision exercise.

Evaluation of the revision should stress specific temperatures, times, shapes, and quantities.

Home Made Horse Shoes

Select a piece of soft iron stock, heat it, and quench it in oil. When it has been in the oil bath long enough, it should be removed and sprinkled with borax and then plunged at once into the heat of the forge. In a few minutes when it looks ready, remove it and hammer it hard on the anvil to make the end roundish in shape. When complete, it should be tested on the forefinger to see if it balances. Be careful that the heating has not destroyed the temper too much.

As an alternate assignment, you might ask students to think back to the tissue paper rose and try their hands at rewriting those instructions.

15 Oral Briefing versus Technical Report: Two Approaches to Communication Problems

William E. McCarron
United States Air Force Academy

Reporting orally in the classroom helps students prepare for the numerous informal and formal oral briefings they will deliver in professional life. Class work stresses differences between oral and written technical communication, practice with graphics, adjustment to feedback, and the importance of *ethos*.

Teachers of technical communication are rightly concerned that students write abstracts, descriptions of mechanisms, and reports; however, an important adjunct is frequently shortchanged: the oral reporting of technical information. This paper outlines key differences between the written technical report and the oral technical briefing, beginning with obvious differences and proceeding to subtle but important distinctions.

Before the analysis of the differences between written and verbal technical communication, a brief discussion of the demand for oral technical skills may prove helpful. Harold P. Erickson reports that technicians employed by private companies devote twenty-five percent of their time to oral reporting: "The interviews [in private companies] definitely showed oral reporting to be a major factor in research work. It is nearly impossible to separate conversational reporting from factual reporting to a superior or subordinate."[1]

As a result, Coleman Technical Institute, at which Erickson is an instructor, developed a technical writing curriculum centering around oral communication. Moving in a similar direction, John A. Walter, premier teacher of technical writing at the University of Texas at Austin, sets aside at least two weeks of every semester for students to report orally on aspects of their research.[2] My own experience as an instructor at the United States Air Force Academy has convinced me to require students to demonstrate mastery of technical information by giving a ten-minute, stand-up briefing to an audience unfamiliar with the subject matter. Students brief fellow students who are majoring in various science and engineering fields, and each briefing, therefore, must convey technical information to an audience that is largely unaware of the intricacies of the briefer's field. In fact, every science and engineering major at the Academy is required to take technical writing, and twenty percent of his or her grade depends on the ability shown in two oral briefings during the course of the semester.

As a final introductory note, many of the examples used here come from the world of business and industry. Students of technical writing may be hired for their science and engineering talents, but they will also be expected to write and brief on these scientific and engineering subjects. In fact, amid the controversy over getting back to basics, I am pleased to see English teachers expressing concern about finding out what industry wants from writers. In a recent issue of *College English*, Joseph M. Williams voiced that concern: "Even worse, we know almost nothing about the way individuals judge the quality of writing in places like Sears and General Motors and Quaker Oats. What counts as good writing at Exxon?"[3]

What counts for good writing at Exxon is clear, crisp technical writing, whether it be an oil geologist's site survey running to a hundred pages or the president's annual report to stockholders running a mere four or five pages. By the way, most stockholders' reports are usually first delivered orally to a group of common shareholders whose interests are financial rather than technical.

This brings me to the point at hand—the written technical report and the oral briefing and the differences between them. I will discuss how writers and briefers vary their approaches to cover six areas of common ground: graphics, delivery, organization, feedback, persuasion, and compatibility.

Graphics

One of the most obvious differences between written and oral communication is graphics. Every worthwhile technical writing textbook has a chapter or two on graphics; however, most texts concentrate on the fundamentals of visual aids: labeling of tables and figures, simplifying the sketch of an exploded drawing, creating an orderly block diagram. This approach is reasonable because, whether in a student report or an industrial proposal, the process is still a black and white, two-dimensional one. Color photographs and meticulous illustrations are usually too expensive for the average technical report.

Not so for the oral technical briefing. Multicolored viewgraphs and 35mm color slides are commonplace in even small briefings for business and industry. Most universities and businesses today have a small audiovisual shop that can produce viewgraphs quickly and economically. Even in classroom situations, the most successful briefings come from students who use an overhead or slide projector. Three or four carefully constructed viewgraphs or slides, even if produced in a self-help audiovisual shop, are more convincing than a flipchart where diagrams are quickly drawn with felt-tipped pens. I am not trying to underplay the use of simple visual aids; I am simply pointing out that business and industry use viewgraphs and slides as a matter of course, and the closer the classroom teacher can simulate the setup of a small conference room, the better the correlation between student efforts and professional expectations. If graphics equipment is not available in the classroom, teachers can arrange a demonstration by the college audiovisual shop, or, even better, ask an experienced technical briefer from the business community to give a fifteen-minute briefing in the classroom.

In terms of importance, good graphics can be to the oral briefing what a detailed outline is to the written report. Well-constructed viewgraphs outline for the viewer the key points the briefer is making. The briefer, much more obviously than the writer, appeals to sight and sound. Because of this appeal, the visual aid should contain in abbreviated form the key points the briefer will discuss in detail. In fact, when I am preparing a technical briefing for oral delivery, I keep two pads of paper in front of me; on one I write the text and on the other I sketch an "outline" (ideas for diagrams or tables) that will visually establish the points I intend to make.

In report writing, on the other hand, a graph or diagram is an adjunct to the written explanation. That is, the graph or diagram must be related to the written argument and carefully explained in that document. For example, test reports on the Air Force's new F-16 fighter aircraft usually begin with a technician's drawing of the F-16 and a detailed written description of the aircraft. The drawing is dramatic, but the technical writing describes the parts of the system and how they work. In an oral briefing on the same subject, the briefer relies on a pointer and a detailed sequence of visual aids. The briefer literally points to the various parts of the F-16 while discussing each.

By way of analogy to literary genres, the technical report is a novel with occasional illustrations that support the novelist's descriptions. The briefing, however, is a drama in which the briefer is an actor who speaks lines and depends on movement and stage props to communicate with the audience. Or, viewed mathematically, the written report is two-dimensional with words and diagrams forever committed to the printed page. In contrast, the interaction of briefer with visual aids is three-dimensional and, quite literally, colorful if the briefer uses color-coded visual aids and a bit of voice-enthusiasm for the subject. In short, visual aids are equal in importance to the words the briefer utters. If members of the audience do not always listen to what the briefer says, they nevertheless continue to see the important points on the screen in front of them. And, by the way, the soundest compliment that can be paid to good graphics occurs when a briefing is over. If several members of the audience ask for photocopies of the briefing slides rather than for a copy of the script, the briefer knows he or she has created self-explanatory visuals.

An ever-expanding audiovisual technology continues to improve the graphic aids available to the technical briefer. Many conference rooms feature reverse projection screens with the audiovisual equipment unseen behind a smoke-colored screen and the images projected through the translucent glass screen. The viewing experience is like that of watching a movie with no projection camera in sight, and the briefer, free from the clutter of cords and portable screens, simply presses a button on the podium and an out-of-sight assistant puts on a new viewgraph. Sony, 3M, and a host of other companies now produce such graphic aids as portable cassette videotape recorders, which can supplement an over-the-table or formal briefing. Even though college students cannot be expected to master sophisticated graphics aids, they should be familiar with them because these are the devices they will use in their future work.

Delivery

We are all familiar with news presentations on TV, we listen every day to students in our classes, and we have all heard scholarly papers delivered at conventions. A speaker's enthusiasm for the subject, voice modulation, and inflection enliven even the dullest subject. Conversely, even an interesting subject fails to attract listeners when it is delivered in a monotone. Even the most serious depiction of, say, the technical specifications on a new stereo system allows for enthusiasm, even a bit of humor.

Technical writing is a different medium entirely. The author can control tone, the order of the subject matter, and the consistency of the report, but he or she cannot personally enliven the material. Again, as in drama, elocution and delivery are key factors in the oral presentation of technical information. The ability of the briefer to speak with conviction and to interact with the audience and the visual aids (the background scenery) is vital to the success of the oral briefing.

Organization

Whether students are writing a description of a fingernail clipper or a detailed explanation of computer software used to support an electronic communication system, the principle is the same: put the purpose statement up front and guide the reader to an overall view of the subject. If the budding technical writer fails to provide a thesis statement at the beginning of the report, the judicious reader will flip forward to get a general view of the subject matter. Chances are, there is a summary somewhere in the report. In short, the technical writer may "blow" an explanation or transition in a report, but the reader who has the leisure (or the patience) to sift out the important information will still garner the message.

Such is not the case in the oral technical briefing. To ignore a clear thesis statement or overview at the outset spells disaster because the audience does not have the opportunity to skip ahead and pick up key points. Of course organization in a written report is an important element and devoutly to be hoped for, but organization in the time-bound technical briefing is absolutely crucial. Not only is the oral briefer obligated to lead the audience through each point in an orderly manner, but he or she has a limited time in which to do so.

Few people remember Edward Everett, the renowned orator who preceded Lincoln at Gettysburg. They remember Lincoln, who spoke briefly and poignantly. Even the most detailed technical briefings rarely last longer than thirty or forty minutes; an audience simply cannot absorb information after that period of time. Again, as was the case with graphics, the oral briefing is a more dramatic genre than is the written report. The briefing is a dramatic monologue in one act, and the briefer must make his or her points in one concise stage appearance. The reader of a lengthy technical report can, as with a novel, put the report down and return to it at a later time.

Every technical writer must be selective about subject matter. In describing how a microcomputer works, the technical writer can rarely dwell on the function of every microchip in the device but must choose key components and discuss how they interact. If selectivity is important for the technical writer, it is essential for the technical briefer. First, the briefer lacks the luxury of time in front of an audience; and second, oral briefing is a slower process than silent reading. The consequence is obvious: the briefer can give a broad, general treatment of microchips or he or she can discuss one or two aspects in detail, leaving the listener with the impression that, were time available, the discussion of all aspects would be just as thorough.

A concrete example might illustrate the importance of selectivity in organizing the oral briefing. The Air Force Test and Evaluation Center recently completed a lengthy study of the F-4G Wild Weasel, an electronically equipped fighter aircraft whose mission is to seek out and destroy enemy ground radar sites. The written report centered on tests conducted against some sixteen test objectives, ranging from engine performance to maintenance costs for spare parts. Department of Defense officials who read the report, and whose decision it ultimately was to approve or disapprove full production of the system, had a complete survey of the F-4G's performance. Prior to the completion of F-4G testing, however, the Pentagon wanted a briefing on how the testing was going. The question was a simple one: "Were the critical test objectives being met?" Enter the oral technical briefer in the person of the F-4G test director. Obviously, he could not give a complete history of all testing conducted to date. He had to be selective, and he gave his questioners what they asked for—a preliminary briefing on several of the most important areas of testing backed up with clear, precise graphics.

This distinction between written and oral technical communication is important for students, teachers, and business people alike. With congressional committees staring into the faces of personnel in the defense industry and with stockholders wary of management's new ventures, the convincer is not likely to be a detailed technical report but a face-to-face explanation before a live audience. The oral communication of selected data persuades an anxious audience, as long as that information is supported by a detailed technical report.

Closely related to the principle of selectivity is the concept of frequency. One technical report on a subject will often suffice, but its highpoints are apt to be communicated orally many times to a variety of audiences. I have known the same basic technical material to be discussed informally over the desk to a corporation president, briefed semiformally to a corporation staff, and delivered formally to a meeting of expert devil's advocates from outside the corporation.

Earlier in this article, I cited Professor Erickson's observation that a quarter of a technical expert's time is involved with oral briefing. I am willing to wager that with the advent of new technology the percentage of time has increased in the last few years. Aside from the actual research or data gathering involved, the demand in industry today is for informal and formal briefings on the results of research and experiment. The greatest breakthroughs are ineffectual unless they are conveyed to decisionmakers. David M. McLean, a technical writer and briefer at Martin Marietta's Orlando, Florida, Division, is right in insisting that writers and editors must be able to handle themselves on their feet with all levels of personnel.[4] McLean's point is that the oral communication of technical data is both important and frequent. Following McLean's cue, I ask each of my students during the time set aside for the writing of major reports to come to my office to brief me on how the research is progressing. I follow this informal update with a full-fledged classroom briefing several weeks later in which each student formally briefs fellow classmates on the continuing progress of his or her project. Such a procedure requires more office hours for the technical writing instructor and some sacrifice in classroom contact, but the experience is vital because students will be expected later on in their careers to give frequent, well-organized briefings to a variety of audiences.

Feedback

Feedback is a simple concept. It is the reaction of the reader or listener to what a technician writes or says and the effect of that reaction on the writer or speaker. Feedback to technical writing is usually slow and deliberate; in an oral briefing, it is quick, often instantaneous. The executive reviewing a written report may make marginal notes or compile a list of comments. These comments, however, are generally one-way and often imprecise: "Rewrite this section" or "Add details." The reviewer simply cannot elaborate on every point and, what is more, the originator has to judge the temper of the reviewer before making changes.

On the other hand, the oral briefer, even in the most formal presentation, is the recipient (or victim!) of instant feedback. That feedback can be direct or indirect. Direct feedback is a sudden question from the audience that may interrupt the briefer in mid-sentence. That's right. My experience in the oral communication of technical information has not been one of the "canned" presentation with questions politely withheld until the set speech is completed. More often, the audience interrupts at any time to ask a relevant question.

A military officer I know, for example, recently briefed members of the federal government's Office of Management and Budget on how the Air Force tests and evaluates new military hardware. Not three minutes into the briefing, he was interrupted by a budget analyst who shouted out this question: "How cost-effective is the program you are describing?" The briefer handled the question well. More important, he recognized the feedback, adjusted his presentation slightly, and specifically mentioned budget and cost aspects during the remainder of the briefing, even though budget and cost were not central to the briefing he was delivering.

Audience analysis and reaction to feedback are essential concerns for every technical briefer. Not only must the briefer react to direct feedback, but he or she must be sensitive to indirect feedback, another area in which the oral briefing differs dramatically from the written report. A lifted eyebrow, a yawn, a laugh are forms of indirect feedback to the oral briefer. The closest approximation to this type of feedback to the written report is a phone call from the boss's secretary informing the anxious writer that the boss fell asleep at page three of the report.

What is more, indirect feedback is more complex than the audience's reaction to what it hears the briefer say. An audience is just as likely to react to the visual aids a briefer uses as to the words he or she speaks. Thus, the technical briefer must respond to feedback on what an audience sees as well as on what it hears. A puzzled look may result from what a member of the audience sees on the briefing screen or because that person is unable to relate what he or she sees on the screen to the remarks the briefer is making. The more skilled the oral briefer, the more apt he or she will be to interpret this indirect feedback and to make adjustments in the presentation.

To continue the analogy with drama, the greatest play suffers if the actors cannot adjust themselves to the reactions of the audience. Pace and timing, even improvisation, are essential to a successful performance. So, too, with the technical briefer. If the audience seems disinterested in a particular point, the briefer had better change the strategy. Many a technical briefer has thought, "Oh, oh, I'd better skip the next slide or two and make a skillful transition to another aspect of the subject because my audience isn't really interested in these points."

Certainly, the objective communication of technical information is uppermost in the mind of the technical briefer, but the audience is not always a group of fellow professionals steeped in a knowledge of the material being presented. For a given technical briefing, the audience may initially be engineers, then middle managers in the corporation, then the city council who must vote on a technical proposal. The briefer-actor must adapt to these audiences and respond to the feedback they generate.

One final difference between report and briefing feedback needs to be emphasized. When a report has been typed or printed in final form, the writer is locked in. If vital new material must be added, the report must be recalled (and I have seen this happen with industrial reports), and an errata sheet added or an annex printed. The oral briefing, on the other hand, is as dynamic as the Dow-Jones Industrial Average. It can be modified or expanded as necessary to meet the needs of a variety of audiences as long as the central content of the briefing remains the same.

Persuasion

At a recent Rocky Mountain Modern Language Association meeting, I had the pleasure of serving as moderator on a technical writing panel. Panelist Marion K. Smith, a veteran technical writing teacher at Brigham Young University, emphasized that the intent of technical writing is not merely to inform readers about the subject and certainly not to impress them with the author's knowledge of an intricate technical subject.[5] Instead, solid technical writing persuades readers to accept particular points about a subject. Thus, technical writing fits solidly into the tradition of classical rhetoric: it is language and argument calculated to persuade an audience, for example, that one technical proposal for a sewage treatment plant is more beneficial than another.

The same principle can be carried over to the oral briefing, but the persuasive marshalling of technical arguments is even more important to the briefer. First, the briefer must select the arguments that best support key points. Second, the briefer must make those arguments convincing to a live audience in a limited amount of time. Finally—and this is perhaps the most important point—the success of technical briefing often depends not so much on the quality of the technical arguments as on the integrity and stature of the briefer.

This last point deserves explanation. In industry, for example, a technical report on a satellite communications system may run several hundred pages and contain input from a dozen writers. Each writer's contribution must be carefully integrated into a comprehensive report that stands on the merits of its own arguments. An oral briefing on that same subject is apt to be given by a high-ranking executive who has the brief written for him or her and tailored to his or her personality and style. The briefing on satellite communications has force and conviction because of its arguments, of course, but it has added weight precisely because a knowledgeable, important executive is giving it. In the armed forces, it is a general officer who briefs a congressional committee on a research project. In industry, it is the bank president who briefs the chamber of commerce on the dollars to be loaned for a new civic project. In other words, the author of a technical report is seldom evident to the reader; in technical briefing, however, the briefer stands up and endorses the technical project or proposal, putting his or her reputation directly on the line.

In fact, the integrity of the briefer—the audience's belief that the briefer will not distort the subject matter—has traditionally been the single strongest argument in public communication. Aristotle, Cicero, and Quintilian all note that *ethos* (the character or integrity of the speaker) is the chief argument an orator can bring to a subject.[6] If the orator believes in what he or she is saying, that conviction is carried over to the audience. The two remaining general sources of classical rhetorical argument, *logos* (the arguments from reason and evidence) and *pathos* (the emotional appeal the speaker brings to the subject through word and action), are essential, too. But the primacy of *ethos* is never more evident than in oral briefing, whether the subject is political, literary, or technical.

In the world of technical communication, the importance of *ethos* is everywhere evident. Most industries have a staff of speech writers who translate the technician's laboratory results into clear information that a senior manager can use to brief a variety of audiences, both technical and nontechnical. Sales companies, for example, bombard me with brochures and detailed specifications on new copier and reproduction equipment for my office. Then, a sales representative visits my office and talks about important features of the new equipment. If I am still dubious, the representative invites me to the home office for a demonstration of the equipment, even a talk with the boss on the special benefits of the equipment. In other words, as the technical information being conveyed becomes more important, so, too, does the stature of the person transmitting it.

Technical briefing is frequently more persuasive than the technical report in at least one other way. In the business world, the oral briefing is more important and persuasive simply because it is available sooner than the finished technical report. An expert technical briefer can put together a comprehensive briefing in several days, whereas the written report, even in draft, requires time-consuming and laborious writing and editing. Note that I use the words "time-consuming" and "laborious" with respect. The preliminary technical briefing, for all its importance and speaker conviction, is no substitute for the detailed technical report that delineates all final conclusions in thorough, persuasive prose.

Compatibility

Because the technical briefing so often precedes the formal technical report, one final point needs emphasis: compatibility. Compatibility means that the oral briefing and the technical report must not contradict each other. Since the final written report might well differ from the preliminary briefing in some of its conclusions, the writer must remind the readers precisely where the completed report differs from the briefing.

In my experience, I have seen technical writers confronted with objections of this sort: "Two conclusions in your report differ from what you said two months ago." Carefully placed transitions in the final report can prevent this type of objection. So, too, can the briefer's words: "I caution you that these results are preliminary. The written report will contain the final conclusions and recommendations on the project."

Compatibility is particularly important when the technical briefer and technical writer are different people, but compatibility is usually assured when the briefer and writer sit down across the table and discuss how each will treat the subject. They make sure that terminology, emphasis, and organization parallel each other in the two forms of technical communication.

Conclusion

The discussion of differences between the technical briefing and the technical report is, of course, somewhat artificial. In actual experience, the six distinctions I have made between briefing and writing overlap each other. One cannot, for example, develop graphics for a briefing without realizing that the organization and delivery of the briefing differ significantly from the presentation in the final written report. Still, teachers of technical writing do students a disservice unless they stress the differences between briefing and report writing. Occasionally, this will mean going outside the classroom to observe the demands that business and industry are now making on the technical briefer and writer.

Technology today is pointed toward increased oral communication, and innovations such as electronic mail, laser communications, and teleconferencing are on the verge of becoming business practice. A few years from now a briefer's presentation will be beamed via satellite to company audiences or to audiences in other parts of the world. The young technician's knowledge of graphics and his or her ability to organize ideas, interpret feedback, and persuade by the force of his or her subject matter and personality will spell success or failure in the growing world of oral technical communication.

Notes

1. Harold P. Erickson, "English Skills among Technicians in Industry," in *The Teaching of Technical Writing*, ed. Donald H. Cunningham and Herman A. Estrin (Urbana, Ill.: National Council of Teachers of English, 1975), pp. 158–59.

2. John A. Walter, "Confessions of a Teacher of Technical Writing," *The Technical Writing Teacher* 1 (1974): 9.

3. Joseph M. Williams, "Linguistic Responsibility," *College English* 39 (September 1977): 13.

4. David M. McLean, "The Demands of Industry on the Technical Writer," in *The Teaching of Technical Writing*, ed. Cunningham and Estrin, pp. 149-50.

5. Marion K. Smith, "Observations of a Veteran Tech Writing Teacher," address delivered to the Rocky Mountain Modern Language Association, October 22, 1977.

6. Since most classical rhetorical theory deals with oral rather than written communication, the technical writing teacher is well advised to look into Aristotle's *Rhetoric*, Cicero's *De Oratore*, and Quintilian's *Institutio Oratoria;* see also, Andrea A. Lunsford, "Classical Rhetoric and Technical Writing," *College Composition and Communication* 27 (October 1976): 289-91.

Part Three: Exercises

16 Technical Writing Class: Day One

Dean G. Hall
Wayne State University

This entertaining plan for the first class meeting captures the interest of students while convincing them of the importance of learning to write competently in their technical fields.

I have found that most students from technical fields returning from the meaty courses of their various majors to take a technical writing course in the English department have the notion (perhaps well founded) that the course will be of little use to them. Their reaction to the first meeting of their technical writing class can, therefore, predict success or failure for an entire term. If the instructor makes a positive impression on the first day, he or she can require many unexciting but necessary activities for the remainder of the term and students will tend to follow along, trusting that the material will be shown to be relevant to their needs. If, on the other hand, the instructor fails to convince students of the viability of the course at its outset, he or she may not get them interested for the rest of the term. Realizing, then, the importance of an immediately positive image, I have overcome my reluctance to share my opening day activity. I hesitated because the idea is so simple and because it runs counter to the rubric that the tone of a technical writing class must at all times be "professional." What follows, then, is a summary of my slightly unorthodox opening day in technical writing class.

After verifying that the folks in front of me are indeed the tech writing class and after briefly introducing myself, I ask them to remove everything from the tops of their desks except three pieces of paper. When the noise of ripping paper and sliding books abates, I tell them, "Make a paper airplane from one of the pieces of paper." This request is usually met by grins, groans, head-shaking, or craning of necks to see if anyone else has heard the same instruction. I allow as much time as students need, but a minute or two usually suffices. I also tell students to put their names on their finished products. By now, though reasonably confident of their aircraft, students are usually apprehensive about the activity's relevance to their careers both within and outside the class. I now encourage a practice flight. As twenty-five planes soar and land around the room, the need for owner identification becomes obvious. I then ask students to retrieve their planes. The class is now loosened up a bit, for the motion and activity force them to notice each other—though they still look at me as if I were out of touch with the harsh realities of their chosen disciplines.

Then I announce, "Take the second piece of paper and write instructions for someone to recreate exactly the plane you have just built and flown. Sign your instructions." Students nearly always ask if they can draw pictures; I tell them to include whatever is necessary to complete the assignment. Though I say that students can take as long as they need, after fifteen minutes I begin to prod them about finishing. I encourage students to note how much longer the writing takes than the actual construction did.

When all have finished, I ask them to exchange instructions with someone in front of or behind them; and the light goes on for a few as they realize what the third piece of paper is for. I now say, "This is the first and last time in the whole term that I want you to be stupid. I want you to follow the instructions in front of you, but I do not want you to supply any gaps in reasoning or procedure for the writer. Do the writer no favors. And most important, if you cannot go from one step to the next without supplying a step that the writer has not included, just quit." After this statement, I allow a few minutes for the construction of the second planes.

Each student now returns the instructions to the author and compares the initial plane with the second one. This encounter allows each student to meet at least one other person and is usually accompanied by snickers or gasps of disbelief as students try to explain why they could not follow given instructions or what they had really meant to say. (Only once have I had both parties in an exchange produce duplicate planes. I first introduced the activity with aerospace students—probably the perfect audience

—but I have since found it effective with students from all disciplines.) By now the class is quite open, and by moving among the students I bring to their attention the most outrageous failures, thereby augmenting their perception that something can go dreadfully wrong in the writing process.

Now comes the crucial part of the class period. Though students realize on an as-yet-unarticulated level what has just happened, the perception needs to be reinforced. Here is an excellent opportunity to make clear key concepts of the course. Students are receptive in this situation since their inadequacies were pointed out as much by themselves as by me. I offer, usually through a simple list on the board, three cardinal rules of technical writing.

1. Producing a good piece of writing takes time and may be just as difficult as building a product. Since most students are quantitatively oriented, by computing at the board the ratio of the doing time to the writing time in the airplane example (usually a factor of at least ten to one), I can make this truth stay with them.

2. A writer cannot assume that the audience knows exactly what is in his or her head, not even when that audience is made up of the writer's colleagues. What was obvious to each airplane builder was certainly not obvious to the person who tried to follow the instructions. And I add that this problem will actually intensify as students acquire more expertise in their respective fields.

3. Technical writing that allows more than one interpretation is unacceptable. Therefore, I give students this statement as a general evaluative tool: "If the writing you do for this class can be understood in more than one way, it is wrong."

In this way, three important considerations in technical writing —adequate preparation time, audience analysis, and precision—are introduced and made concrete during the first class meeting. Students have their planes and instructions in front of them as evidence. In addition to the concepts about technical writing that the first class makes clear, there are some definite psychological advantages to the session. Students share an experience, one that lets them know their problems are not theirs alone. They feel comfortable in the class since they have talked to at least one other person, and their own observations and deductions endorse those made by the instructor.

I usually close with several examples, from my own experience or from texts, of writing practices that have proved dangerous or expensive or both. I impress upon students that they are the designers and makers of the future and that their writing is a primary way to meet their responsibilities for safe and efficient production. Obviously no one was killed in a paper plane, but sloppy writing in the real world does irreparable harm. I have found that students take the course seriously after this first-day activity, even though the original class was entertaining. Nearly all the students will stay—and they will do the work.

17 Organizing Is Not Enough!

Paul V. Anderson
Miami University

Technical writing classes obviously must teach students to write
organized prose; however, students also need to learn how to
reveal that organization to their readers through content signals
and through the arrangement of material on the page. How to
develop these skills in two fifty-minute class periods is discussed.

"Industry and frugality," Benjamin Franklin tells us in his *Auto-biography*, are "the Means of procuring Wealth and thereby securing Virtue." Franklin adds, however, that one cannot gain those objectives simply by being industrious and frugal "in *Reality*"; one must also "avoid all *Appearances* of the Contrary." Accordingly, while Franklin was building his small printing shop into a flourishing publishing company, he never allowed himself to be seen in "Places of idle Diversion," he never went a-fishing or shooting, and when he allowed a book to seduce him from his labor, he did so privately, to prevent scandal. Franklin was rewarded for his efforts. At forty-two, he had become wealthy enough to retire from business to devote his full energies to being a statesman, inventor, and man about Europe.

As teachers of technical writing, we can help students see that Franklin's advice applies not only to publishers but also to authors, especially when those authors are pursuing one of the chief virtues of technical writing: careful organization. We must warn students that no matter how industriously they organize their material, they must also strive to make that organization readily apparent to their readers.

Combining suggestions from others with some ideas of my own, I have developed materials for two fifty-minute class periods in which I teach students how to reveal the organization of their writing.[1] The strategy may be used with students from high

Paul V. Anderson

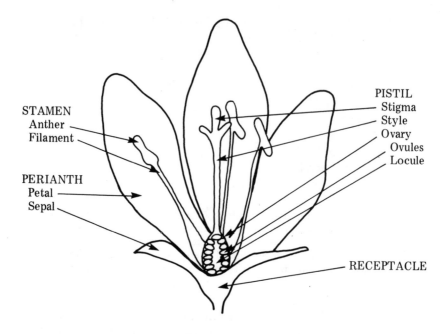

Author's View	Possible View of Reader	Preferable View of Reader
FLOWER	FLOWER	FLOWER
STAMEN	STAMEN	I. STAMEN
Anther	ANTHER	A. Anther
Filament	FILAMENT	B. Filament
PISTIL	PISTIL	II. PISTIL
Stigma	STIGMA	A. Stigma
Style	STYLE	B. Style
Ovary	OVARY	C. Ovary
Ovules	OVULES	1. Ovules
Locule	LOCULE	2. Locule
PERIANTH	PERIANTH	III. PERIANTH
Petals	PETALS	A. Petals
Sepals	SEPALS	B. Sepals
RECEPTACLE	RECEPTACLE	IV. RECEPTACLE

Figure 1. Diagram for explaining the difference between author and reader views of the organization of a written communication.

school through graduate school, although the teaching aids reproduced here were designed for college seniors and graduate students and for professional researchers and engineers. Using these materials, I first explain the need to signal organization to readers, and I then introduce techniques that can be used to meet that need.

Establishing the Need for Organizational Signals

Most students have never thought about the need to reveal organization. Like most authors, they usually think of their writing from their own point of view, not from that of their readers. In fact, many students are only vaguely aware that there are two points of view to be considered. They simply presume that an organization that is evident to them will be equally evident to their readers.

To explain the difference between the way authors see the organization of their writing and the way readers see it, I show an overhead transparency of the illustration in Figure 1. At the top is a diagram of a flower that represents an author's view of the information to be conveyed to a reader. At the bottom are three panels, each representing a different view of what the author has written about the subject. Initially, I cover the bottom three panels so that only the labeled diagram is exposed. I then uncover the bottom panels, one at a time, as I explain the differences between the views of the author and the reader.

I begin with the author's view, which is represented in the left-hand panel. Authors see their organization as a pattern with two dimensions: logic and sequence. Authors see, first, the logical relationships of coordination and subordination that they have established among the various parts of their discussion. In outlines they would indicate these relationships through indentation; therefore, I have indicated this dimension of organization with a horizontal arrow. Authors also see a second dimension to their organization: the overall sequence of the parts of their material. In their outlines authors would indicate this sequence by the order in which they list the parts of their discussion; accordingly I have represented the second dimension of organization with a vertical arrow.

A reader's view of a document's organization is often more limited than an author's two-dimensional view because reading, by its very nature, is a one-dimensional activity. That is, readers read linearly, one section after another, one paragraph after another, one word after another. Restricted to reading one element at a time, readers will undoubtedly learn the sequence of the elements of a document—at least of those elements they have encountered so far—but they may fail to perceive the logical relationships of coordination and subordination that an author has established within the material. As a result, readers may see the contents of a document as a mere list, like that shown in the center panel at the bottom of Figure 1.

To help readers gain a two-dimensional view of organization, authors must recognize that a well-written document contains two distinct kinds of information: information about subject matter and information that explains how the discussion of that subject has been organized. The right-hand panel at the bottom of Figure 1 shows the relationship between these two. The information about subject matter is represented by the list of topics, which is the same as the list shown in the center panel. This information is what readers read the document to learn. However, if they are to understand that information fully, they need to know how it is organized. They can learn the sequential dimension of that organization simply by reading the document, but they can perceive the logical dimension only if authors add information about the logical relationships of coordination and subordination. In the right-hand panel of Figure 1, these relationships are represented by the outlining apparatus of Roman and Arabic numerals and upper- and lowercase letters. Only by supplementing information about subject matter with such organizational information can authors fully communicate their material to their readers.

To enable students to experience the kind of problem they create for their readers when they neglect to provide such organizational information, I show them an overhead transparency of the list (but not the brackets) given in Figure 2. The list is a poorly designed table of contents for a booklet about the Pegasus Automatic Balance, a fictitious laboratory instrument used to weigh samples of up to one hundred grams. The author of this booklet carefully organized the material but failed to reveal that organization. As a result, readers must try to discover it for themselves—and that is exactly what I ask my students to do.

To guide students in their search, I ask them to identify items that the author seems to have placed next to each other because the items are actually components of larger units of discussion.

Thus, for example, items 1 through 4 belong together because they all deal with knobs, and items 12 and 13 go together because they both have to do with weighing. As students point out each group of items, I enclose the items in a bracket as shown in the left-hand margin of Figure 2. When my students seem unable to find more groups, I ask them to look at the groups they have already identified to see whether some of those groups have subgroups. In response, students point out, for example, that the group about knobs, mentioned above, contains a subgroup—items 1 and 2, the weight-application knobs. Similarly, I ask students to consider whether some of the groups they have found are actually subgroups of more comprehensive units; for example, the "weighing" group—items 12 and 13—belongs with the "adjusting" group—items 10 and 11—to form a group on the operation of the balance, a group entirely distinct from the group formed by the first nine items, all of which deal with the construction of the mechanism.

After students have requested about as many brackets as are shown in the margin of Figure 2, I display a transparency showing a revised table of contents (Figure 3). Students are invariably surprised to discover how much easier the revised version is to read, even though the only difference is the addition of organizational signals. After asking two purely rhetorical questions —which table would students rather encounter as readers, and which would they rather provide their readers when they themselves are authors—I promise to spend the rest of the class session, and all of the next, teaching the skills used to provide organizational information to readers.

Providing Organizational Information

As I explain to students, they can provide organizational information to their readers by adding material to their written communications and by purposefully adjusting the way material looks on the page.

Adding Material

Students—and other authors—can add four kinds of material to their prose to reveal its organization: headings and subheadings,

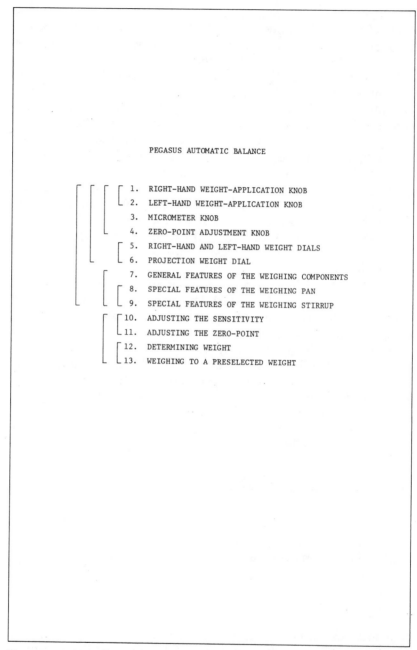

Figure 2. Table of contents to which the author has neglected to add organizational information.

PEGASUS AUTOMATIC BALANCE

I.   CONSTRUCTION

A.   CONTROL COMPONENTS
    1.   KNOBS
        a.   Weight-Application Knobs
            Right-hand Weight-Application Knob
            Left-hand Weight-Application Knob
        b.   Micrometer Knob
        c.   Zero-Point Adjustment Knob
    2.   DIALS
        a.   Right-hand and Left-hand Weight Dials
        b.   Projection Weight Dial
B.   WEIGHING COMPONENTS
    1.   GENERAL FEATURES OF THE WEIGHING COMPONENTS
    2.   SPECIAL FEATURES OF THE WEIGHING COMPONENTS
        a.   Special Features of the Weighing Pan
        b.   Special Features of the Weighing Stirrup

II.   OPERATION

A.   PRELIMINARY ADJUSTMENTS
    1.   ADJUSTING SENSITIVITY
    2.   ADJUSTING THE ZERO-POINT
B.   WEIGHING OPERATIONS
    1.   DETERMINING WEIGHT
    2.   WEIGHING TO A PRESELECTED WEIGHT

Figure 3. Table of contents to which the author has added organizational information.

B. PRODUCT HANDLING

In our guidelines on processing, we now have completed our suggestions for guarding against microbial contamination. We turn now to our guidelines on how to handle the product. Here, we consider the receiving of the new raw material and the processing of it.

1. Receiving Raw Materials

By raw materials, we mean both the fish and any other raw materials used in processing.

a. Fish.—We consider first the fresh fish and then the frozen fishery products.

(1) Fresh fish.

Check fresh fish for sign of spoilage, off odors, and damage upon their arrival at your plant. Discard any spoiled fish.

Immediately move fresh fish under cover to prevent contamination by insects, sea gulls, other birds, and rodents. If the fish are to be scaled, scale them before you wash them.

Unload the fish immediately into a washing tank. Use potable, nonrecirculated water containing 20 parts per million of available chlorine and chill to 40°F or lower. Spray wash the fish with chlorinated water after taking them from the wash tank (Fig. 11).

If incoming fresh fish cannot be processed immediately, inspect them, cull out the spoiled fish, and re-ice the acceptable fish in clean boxes; then store them preferably in a cold room at 32° to 40°F or, at least, in an area protected from the sun and weather and from insects and vermin. Wash, rinse, and steam-clean carts, boxes, barrels, and trucks used to transport the fresh fish to the plant if any of these are to be used again. Reusable containers should be rinsed again with chlorinated or potable water just before use. *Note:* wooden boxes and barrels should not be reused. It is virtualy impossible to satisfactorily sanitize used wooden containers such as fish boxes and barrels. If disposable-type containers are used, rinse them off and store them in a screened area until you remove them from the premises.

(2) Frozen fishery products.

Use a loading zone that provides direct access to a refrigerated room.

32

Figure 4. Page in which the author has carefully revealed the organization of his material to his readers. From J. Perry Lane, *Sanitation Recommendations for Fresh and Frozen Fish Plants* (Seattle: U.S. Department of Commerce, 1974).

numbers and letters, forecasting statements, and transitional statements. Students need no special explanation of the first two kinds of material, especially after discussing Figure 3. In contrast, they do need careful explanations of what forecasting and transitional statements are and do.

A forecasting statement stands at the beginning of a segment of a document and describes the topics to be taken up in the major parts of that segment. Thus, the introduction at the beginning of a long report will contain a forecasting statement that identifies the subject of each of the subsequent chapters in the body of that report. Similarly, the introduction to each of those chapters will contain a forecasting statement that describes the major sections of the chapter, and the major sections will each contain a forecasting statement that tells readers what topics the subsections will address.

Whereas forecasting statements identify the topics an author will treat in the segments of a report, transitional statements signal a shift from one segment to another. Like forecasting statements, therefore, transitional statements appear at every level within the organization of a document: between chapters, between major sections of each chapter, between subsections of those sections, and so on.

In order to show students how to use forecasting and transitional statements, I ask them to study the passage shown in Figure 4, a reproduction of page thirty-two of a government document entitled *Sanitation Recommendations for Fresh and Frozen Fish Plants*. After students have had a moment to glance over that page, I ask them to use the information it provides to outline the section of the book in which this page appears. I begin by pointing out that the first line informs us that the page contains part of a subsection "B" of some major section in the book, one designated by a Roman numeral "I" or "II." "What is the title of the overall section?" I ask. "And what is the title of the previous subsection, called 'A'?" Having read the transitional sentence at the top of the page, students quickly point out that the overall section is "Guidelines on Processing," and that the previous subsection contains suggestions for "Guarding against Microbial Contamination." Similarly, by reading the several forecasting statements on the page, students are able to find other clues to the author's organization. As students call out each detail of the author's outline, I ask them to read the sentence in which they found their evidence, and I transcribe their entry onto an overhead transparency. When the class has finished its work on this exercise, the outline on the transparency looks like the one shown in Figure 5.

Before concluding my remarks on forecasting statements, I address one question that often troubles students: How detailed should a forecast be? "At the beginning of our documents," they often ask, "should we give our entire outline?" The answer, of course, is no. A forecasting statement merely tells what the major divisions of a particular segment are. If those divisions are themselves divided, then each of those divisions will have its own forecasting statement. The author of the page shown in Figure 4 has used forecasting statements correctly. For example, when he announces the major divisions of his advice about "Receiving Raw Materials," he reports that his discussion of that subject contains two parts, one on "Fish" and the other on "Other Raw Materials," but he does not tell what the subdivisions of the part on "Fish" will be. Instead, he correctly places his forecast of those subdivisions within (not before) the part on "Fish."

> II. Guidelines on Processing
>   A. Guarding against Microbial Contamination
>   B. Product Handling
>   1. Receiving Raw Materials
>   a. Fish
>   (1) Fresh Fish
>   (2) Frozen Fishery Products
>   b. Other Raw Materials
>   2. Processing Raw Materials

Figure 5. Outline derived from the forecasting and transitional statements provided in Figure 4.

Up to this point, I have discussed forecasting and transitional statements in the context of the larger units of a document. I have found, however, that no matter how well students understand this material, they need a special exercise to show them how forecasting and transitional statements work within a paragraph. For this exercise, I distribute copies of the rough-draft paragraph shown in Figure 6. Although the author of this paragraph has organized his material carefully, like the author of the table of contents in Figure 2 he has neglected to reveal that organization to the readers, who must, therefore, try to discover that organization for themselves. After letting students guess at that organization for a few minutes, I hand out the author's revision of his rough draft (Figure 7). Then, I ask the class to look for the forecasting and transitional elements within the revision. However, I do not let the students finish by merely telling me that the second sentence is a forecasting statement, or that the third, seventh, and tenth sentences contain transitional elements. I ask them also to notice that the initial words of the fourth, fifth, sixth, and ninth sentences tell the reader that the subject is *not* being changed.

After students finish their work on this paragraph, I ask them to review what they have learned about revealing organization by adding material to the text. The best review, I have found, is to have students examine a response to the kind of assignment upon which they are currently working (Figure 8). By considering how headings, outlining apparatus, forecasting statements, and transitional statements help readers discern the organization of this document, students not only review the skills I have been teaching but also see how another student has used these skills to solve the very kind of communication problem that now faces them.

By the time students have finished examining Figure 8, we are usually about fifteen minutes into the second of the two class periods that I devote to signaling organization. I usually conclude the first just before students have looked at Figures 6 and 7. Even when that work is left until the next period, I have ample time to introduce a second way to reveal organization.

Adjusting the Appearance of Material on the Page

I have found that it is much easier to teach students how to reveal organization by adjusting material on the page than it is to teach

SELECTIVE HARVESTING TECHNIQUES USED BY COMMERCIAL FISHERMEN

TO AVOID CATCHING FISH OF THE WRONG SPECIES

Most selective harvesting is based on behavioral differences. Commercial fishermen can take advantage of these differences by using their knowledge and experience with the aid of devices such as modern navigational equipment and fishing gear. Salmon gill-netters of Puget Sound avoid, even on foggy and stormy nights, certain areas that are known to have large dogfish concentrations. Puget Sound salmon seiners avoid certain areas where juvenile salmon are taken in large numbers. Otter trawlers working La Perouse Bank use accurate navigational aids such as echosounders and Loran to return to their preferred bottomfish tows about 20 miles offshore, some of which lie within one mile of large concentrations of dogfish. Washington and Oregon fishermen trawling for pink shrimp know they can, by modifying footrope weighting, raise their nets a few inches and reduce the bycatch of eel pouts. They can sometimes, by reducing flotation, lower the headrope height and reduce the bycatch of smelt. Troll fishermen find that many factors contribute to the selective capture of certain species, including geographic location, water and gear depth, trolling speed and type of lure.

Figure 6. Paragraph in which the author has neglected to provide forecasting and transitional statements.

SELECTIVE HARVESTING TECHNIQUES USED BY COMMERCIAL FISHERMEN

TO AVOID CATCHING FISH OF THE WRONG SPECIES

Most selective harvesting is based on behavioral differences. The result of these differences often is either a geographical or vertical distributional variation between species or age groups or a difference in food preferences. Commercial fishermen take advantage of geographic differences by using their knowledge and experience with the aid of devices such as modern navigational equipment. For example, salmon gillnetters of Puget Sound avoid, even on foggy and stormy nights, certain areas that are known to have large dogfish concentrations. Also, Puget Sound salmon seiners avoid certain areas where juvenile salmon are taken in large numbers. Similarly, otter trawlers working La Perouse Bank use accurate navigational aids such as echosounders and Loran to return to their preferred bottomfish tows about 20 miles offshore, some of which lie within one mile of large concentrations of dogfish. On the other hand, Washington and Oregon fishermen who trawl for pink shrimp have learned to take advantage of differences in vertical distribution. They know they can, by modifying footrope weighting, raise their nets a few inches and reduce the bycatch of eel pouts. Similarly, by reducing flotation they can sometimes lower the headrope height and reduce the bycatch of smelt. Troll fishermen find that not only do geographic and vertical differences facilitate the selective capture of certain species but also differences in food preferences make trolling speed and type of lure important.

Figure 7. Paragraph in which the author has provided forecasting and transitional statements.

THE RELAY

A relay is an electromagnetically operated device for switching electrical circuits on and off. The parts of a relay much resemble the parts of an ordinary wallswitch for turning on or off the lights in a room, although the internal parts of many relays are invisible because these relays are hermetically sealed, with only electrical leads showing. However, there is one important difference between a relay and a wall-switch: in place of the mechanical switching lever found in wallswitches, a relay has an armature that is moved by an electromagnet. This paper describes the most basic type of relay, and it includes a brief account of other common types.

BASIC RELAY

The relay described in this paper is shown in Figure 1. It has few major parts, and its operation is simple.

CONSTRUCTION

The basic relay has only four main parts: an electromagnet, an armature, a return spring, and a pair of contacts.

Electromagnet. The electromagnet is a piece of ferromagnetic material, called the core, which has wire wrapped or coiled around it. In the relay being described, the core is a cylindrical piece of iron approximately 2 inches long and 1 inch in diameter. The coil is insulated from the core by a thin sheet of paper. This entire part is coated with shellac. When current flows through the coil, a magnetic field is set up so that the core acts like a magnet, pulling the armature toward itself.

Armature. The armature, or movable arm, is a piece of metal hinged at one end with the other end, called the free end, located above the core of the electromagnet. The armature is constructed from a strip of iron that is 4 inches long, 3/4 inch wide, and 1/8 inch thick. It is the movement of the armature, caused by its attraction to the core of the electromagnet, that is the switching motion in the relay.

Return Spring. Attached to the hinged end of the armature is the return spring. The return spring, which is hooked at both ends, is a helical coil spring about 1 inch long; it resembles the spring of a ballpoint pen. This spring is used to return the armature to the position that the armature had been in before being attracted to the core of the electromagnet.

Contacts. At the free end of the armature is a contact, whose mate is situated so that any movement of the armature toward the coil will bring the two contacts together. These contacts are made of copper, 3/4 inch square and 1/4 inch thick, with lead wires attached to connect them to the external circuit in which the relay is being used. When the free end of the armature moves toward the coil, the contacts touch one another, making an electrical connection.

OPERATION

The relay switches on and off in response to the current in the coil of the electromagnet. When there is current in the coil, the magnetized core draws the armature to itself, bringing the contacts together and thereby completing a circuit. When the current ceases to flow through the coil, the core no longer acts as a magnet, so that the return spring can return the armature to its normal position, separating the contacts and switching the circuit off.

OTHER RELAYS

In the basic relay, described above, the contacts make the connection when current flows through the coil of the electromagnet. In some other relays, this connection is broken when the relay is engaged; this can be accomplished by rearranging the location of the contacts. Also, in many relays more than one set of contacts is used. Nevertheless, the principle of operation for all relays is that used by the basic relay.

Figure 8. Student paper in which organization has been carefully revealed.

them to pursue the same objective by adding material to their texts. After all, there are only two simple guidelines to follow when arranging material:

1. Make major segments and their headings more prominent visually than minor segments and their headings.
2. Give visually parallel presentations to elements that hold parallel positions in the document's organization.

When teaching students techniques they can use to give greater prominence to the more major elements in their documents, I divide the techniques into two categories: those that pertain to headings and those that pertain to prose.

By adjusting the appearance of headings, authors can create relatively elaborate hierarchies among those headings. Since the actual number of levels needed in such a hierarchy is small compared to the number of variations that are theoretically possible, many employers and many technical writing textbooks simply suggest a standard format for this hierarchy. However, I find it worthwhile to discuss three aspects of headings that can be adjusted to signal relative importance.

Type size. Headings printed in larger type appear to be more important than headings printed in smaller type. Thus, authors preparing their final copy on a typewriter should type more major headings in uppercase and more minor headings with only the first letter of each word capitalized.

Location. Headings centered on the page seem more prominent than headings tucked against the margin. Similarly, headings at the margin but given a line of their own seem more prominent than headings at the margin that are on the same line as the first sentence of the section they label.

Underlining. An underlined heading appears to be more prominent than the same heading without underlining.

Rather than show students a format that integrates these three variables to create a hierarchy of headings, I ask them to look at Figures 4 and 8 to see how two authors adjusted the appearance of their headings to create hierarchies of importance among them. Also, I ask students to notice that in Figure 4 the author has used an outlining apparatus to make the hierarchy of headings doubly clear to readers.

In contrast to the many ways that authors can adjust the appearance of headings, there are only two ways they can adjust the appearance of prose to reveal its organization. First, they can adjust the left-hand margin of individual sections, as did the author of the page shown in Figure 4. When doing so, authors need simply remember that blocks of prose that have their margins farther to the left appear more prominent than blocks that have margins further to the right. In addition to adjusting the margins of sections, authors can also adjust the amount of vertical space between sections. For example, they might double-space between the end of one minor segment and the beginning of another, but they might triple-space between major segments. To signal the very largest divisions in a long report, authors can begin each of the major sections on its own page, regardless of where on the previous page the preceding segment ends.

Before completing this discussion, I point out that these various techniques can reinforce one another. Thus, for instance, the author of the page shown in Figure 4 has given his most prominent headings to sections that begin farthest to the left, and he has given less prominent headings to sections that begin farther to the right.

After explaining how authors can give greater visual prominence to the elements that hold higher positions in the organizational hierarchy of their writing, I turn to the need to give visually parallel presentations to organizationally parallel elements. I point out that writers confuse readers when they are not consistent in that way. Thus, for instance, if a writer begins one of the major segments of a report on its own page, he or she should begin all other major segments of that report in the same way. Similarly, a writer should assign headings in the same style to all those segments. So that students can examine papers in which authors have given parallel presentations to organizationally parallel elements, I ask them to look again at Figures 4 and 8.

For the final part of the lesson on techniques that provide visual signals, I have students complete two exercises that not only give them practice at applying those techniques but demonstrate how authors restrict the reader's understanding of material when they fail to arrange it carefully on the page.

I begin the first exercise by asking students to explain the organization of the page shown in Figure 9, a page made virtually incomprehensible by its haphazard appearance. In that page, the author describes the taxonomy of *Anoplopoma fimbria*, a mild, white-fleshed fish sold under the name black cod. However, none of my students knows anything about the fish; most do not even know what taxonomy is—and that is the way I prefer it. As they try to discern the author's organization of the material on this page, I do not want them to use clues in the prose; after all, an author should be able to use visual devices that clarify organization even for a reader who does not understand the subject.

To get students started in their search for the organization of the material in Figure 9, I ask them to guess which line on the page contains the heading that holds the highest place in the author's organizational hierarchy. After two or three guesses, I ask students to vote for the line they favor. In most votes, the line beginning with *"Anoplopoma"* wins, even though the very top line is the correct choice. After I reveal the correct answer, I ask students to explain why that choice is not obvious; they invariably point out that while both headings are underlined, *"Anoplopoma"* begins farther to the left. Having learned that "1.2 Taxonomy" is the most major heading on the page, students quickly guess the answer to my next question by telling me that "1.21 Affinities" is the heading for the first major subsection of the section on taxonomy. However, when I ask them to identify the major divisions of the subsection on affinities, they are again stumped because the layout so effectively disguises the fact that "Suprageneric" and "Generic" are parallel headings and that the entire bottom of the page contains information that belongs under the heading "Generic."

To convince students that careful layout can alleviate the reader's difficulties at discerning organization, I show Figure 10, which contains the author's revision of the page shown in Figure 9. After students have explained how the author has rearranged the material to make the organization more readily apparent, I ask them to revise a page themselves. That page is shown in Figure 11. I prod students to continue to make suggestions until they have touched upon almost all the alterations incorporated in the revised page shown in Figure 12, which I then distribute.

I conclude the lessons on revealing organization—both through adding material to a text and through adjusting the appearance of that text—by asking students to look one last time at the student paper shown in Figure 8. If students follow the guidelines introduced in these two class periods, they should be able to do an equally effective job of letting their readers know how they have organized the materials in their own assignments.

1.2 <u>Taxonomy</u>

    1.21 Affinities

     Suprageneric

  Phylum   Chordata

   Class   Osteichthyes

    Order   Scorpaeniformes

     Family   Anoplopomatidae

      Generic

<u>Anoplopoma</u>, Ayres   Jordan and Evermann (1898)

The generic concept used here is that of Jordan and Evermann (1898: 1861–62):

"Body elongate, little compressed, tapering into a very slender
caudal peduncle; head rather long, the snout somewhat tapering;
mouth terminal, moderate, the lower jaw included; maxillary very
narrow, slipping almost entirely under the preorbital; teeth
moderate, cardiform, those in the lower jaw in a single series
laterally, and in a narrow band in front; upper jaw, vomer, and
palatines each with a band of similar teeth; head entirely scaly;
no supraorbital flap; preopercle unarmed, its membranaceous edge
crenulate; gill membranes joined to the isthmus; body entirely
covered with minute ctenoid scales; lateral line single; dorsals

Figure 9. Page in which the author has obscured the organization by carelessly arranging the material on the page.

1.2 Taxonomy
    1.21 Affinities
      - Suprageneric

Phylum	Chordata
Class	Osteichthyes
Order	Scorpaeniformes
Family	Anoplopomatidae

      - Generic

Anoplopoma, Ayres Jordan and Evermann (1898).

The generic concept used here is that of Jordan and Evermann
(1898: 1861-62): "Body elongate, little compressed, tapering
into a very slender caudal peduncle; head rather long, the
snout somewhat tapering; mouth terminal, moderate, the lower
jaw included; maxillary very narrow, slipping almost entirely
under the preorbital; teeth moderate, cardiform, those in the
lower jaw in a single series laterally, and in a narrow band
in front; upper jaw, vomer, and palatines each with a band of
similar teeth; head entirely scaly; no supraorbital flap;
preopercle unarmed, its membranaceous edge crenulate; gill
membranes joined to the isthmus; body entirely covered with
minute ctenoid scales; lateral line single; dorsals

Figure 10. Page in which the author has clarified the organization by rearranging the material shown in Figure 9.

1.23 Subspecies

No subspecies of A. fimbria have been proposed.

1.24 Standard common names, vernacular names

Official common name: Sablefish (American Fisheries Society, Special Pub. No. 61, 1970, p. 57).

Other common names:

   Blackcod (Schultz and DeLacy, 1936: 76)

   Coalfish (Jordan and Gilbert, 1882: 650)

   Skil (Clemens and Wilby, 1961: 240)

   "Mackerel" (Schultz and DeLacy, 1936: 76)

Vernacular name: Beshow (Jordan and Gilbert 1882, p. 650).

The flesh of sablefish has been marketed under various names including:

   Smoked cod (Frey, 1971, p. 70)

   Candlefish (Frey, 1971, p. 70)

   Butterfish (Frey, 1971, p. 70)

   Seatrout (Phleger, et al., 1970, p. 31)

1.3 Morphology

1.31 External Morphology

-Generalized. Phillips, Clothier, and Fry (1954) examined samples of sablefish that were captured from Cape Spencer, Alaska, to Newport Beach in Southern California. The range of meristic counts within these samples is shown in Table 1.

Figure 11. Page in which the author confuses the reader by carelessly arranging the material.

1.23 Subspecies

   No subspecies of <u>A</u>. <u>fimbria</u> have been proposed.

1.24 Standard common names, vernacular names

   - Official common name

      Sablefish (American Fisheries Society, Special Pub. No. 61, 1970: 57)

   - Other common names

      Blackcod (Schultz and DeLacy, 1936: 76)

      Coalfish (Jordan and Gilbert, 1882: 650)

      Skil (Clemens and Wilby, 1961: 240)

      "Mackerel" (Schultz and DeLacy, 1936: 76)

   - Vernacular name

      Beshow (Jordan and Gilbert, 1882: 650)

   - Names given when marketed (selected samples)

      Smoked cod (Frey, 1971: 70)

      Candlefish (Frey, 1971: 70)

      Butterfish (Frey, 1971: 70)

      Seatrout (Phleger, et al., 1970: 31)

1.3 <u>Morphology</u>

   1.31 External Morphology

     - Generalized

        Phillips, Clothier, and Fry (1954) examined samples of sablefish that were captured from Cape Spencer, Alaska, to Newport Beach in Southern California. The range of meristic counts within these samples is shown in Table 1.

Figure 12. Page in which the author has clarified the organization by rearranging the material shown in Figure 11.

Evaluating Student Work

Although it would be possible to evaluate mastery of the skills used to reveal organization by having students perform editing exercises much like those in Figures 2, 6, 9 and 11, I prefer to grade them upon the skill with which they reveal the organization of their own writing. Thus, every assignment students complete during the term provides an opportunity to evaluate how well they have mastered the techniques I introduced in these two class sessions.

I have found, however, that students learn most efficiently when I teach this material while they are working on an assignment that affords an opportunity to use the techniques under discussion. The description of a mechanism is such an assignment. Because the description requires students to develop, as part of partitioning, a modestly complex outline for even a short paper, this assignment prompts them to give special attention to the problem of revealing that outline to their readers. By coordinating my lessons on revealing organization with their descriptions of a mechanism, I ensure that I will have an immediate assignment in which I will be able to comment in detail upon the skill with which students signal the organization of their writing to their readers.

Of course, in that assignment, as in all others, I also evaluate other aspects of writing. After all, revealing organization is only one of a great many skills that students must acquire. However, it is a crucial skill. And it is one that students learn rather quickly, despite their initial skepticism about its value. Apparently they are persuaded by these two days of study that Franklin's advice does apply to their writing: it is not enough to organize their material; to fully enjoy the rewards for that effort, authors must also give the appearance of having organized their writing.

Notes

1. I wish to thank Ian Ellis and Herb Shippen, whom I met while on a consulting assignment with their employer, the National Marine Fisheries Service. Mr. Ellis wrote both the original and revised drafts of the passage on selective harvesting (Figures 6 and 7), and Mr. Shippen wrote the original and revised drafts of the material on *Anoplopoma fibria* (Figures 9 through 12). Finally, I wish to thank Dr. A. James Challis, Instructional Development Specialist, Miami University, for helping me perfect these class sessions.

18 "Build It Again, Sam": An Instructional Simulation Game

Peter R. Klaver
The University of Michigan

BIAS is a newly developed instructional game designed to make students in technical writing classes aware of the problems associated with communication within complex business and government organizations. This paper, which first appeared in the *Proceedings of the North Central Section of the American Society for Engineering Education*, 1978, describes the objectives, rules, and materials of the game and provides a sample game task.

Technical communication focuses attention on the fact that people write and speak to other people within organizations in order to effect actions and change. And people do this in order to achieve both personal and organizational goals, which often conflict.

When teachers design courses based on this recognition, however, they often run into the problem of student disbelief. Unless students already have on-the-job experience with communication problems and organizational conflicts, they are likely to take a skeptical view of the instructor's claim that organizations are made up of real people with conflicting personal and organizational goals and that these factors play an essential part in the design of technical communications. They believe somewhat naively that "If I do technical tasks well, I'm all right."[1]

To introduce these students to the problems of how personal and organizational conflicts affect job performance, goal achievement, and communication, I have developed an instructional game called "Build It Again, Sam" or *BIAS*.[2] The game simulates the organizational and operational climate of a company. It is a *game* because there is competition among players within teams and between teams, a clear-cut method of determining the winner, and rules for playing. It is *instructional* only in a special sense:

the instruction is in the "affective domain." The specific instructional objectives are receiving (attending), responding, and value organization.[3]

Students are divided into two or more teams representing companies that will compete to design and build a prototype product in order to win a lucrative government contract. The game rules, the physical layout, the paperwork, the specified product, the available materials, and the time limitations are carefully balanced so that players face the kinds of communication and organizational conflicts common to working situations. The design task described in this paper requires that an object be built from Tinker Toys, which are inexpensive, durable, and just unreliable enough to cause problems.

The usual pattern of this game is (1) to introduce the players to each other, to the people directing the game, and to the general purpose of the game (this last point is optional); (2) to play the game; and (3) to debrief. During debriefing, a crucial feature of *BIAS*, players and directors discuss what went on in the game, how each person felt about it, how various features of the game might be changed, and, most important here, how the experience bears on what they will be doing in the technical communications course.

BIAS was originally designed for senior engineering students taking a course in Technical and Professional Writing in Industry, Government, and Business. However, either in its present form or suitably modified, it can be played by groups ranging from upper-level high school students to working adults.

The purpose of *BIAS* is to introduce students to communication problems inherent in working in organizations. Specifically, the game introduces the following problems:

1. Defining one's professional task in terms of organizational goals
2. Coping with the conflict between one's own need for information and the need for information of others
3. Dealing with the conflict between one's personal goals and the goals of the organization
4. Facing the effects of the pressure to cooperate produced by people with different specialties and interests (horizontal conflict)
5. Facing the problems created by different attitudes toward and perceptions of organizational goals (vertical conflict)

How Is *BIAS* Played?

This section describes the game, points out features that contribute to its educational and experimental goals, and comments on particular problems and variations.

General Game Description

From twenty to sixty players may participate. Allow at least one and a half hours for playing the game and a minimum of a half hour for debriefing.

Several competing companies are faced with the need to design and produce a prototype of a new product as quickly as possible. The *winning company* is the one that produces an acceptable product (one that meets specified tests) as inexpensively as possible. Costs are determined by cost per unit and by development costs. Cost per unit is determined by the cost of the material from which the product is made. Development costs are the sum of (1) costs of all material used in developing the product, (2) costs of testing the product, and (3) labor costs, which are determined by salaries and bonuses. The *winning individuals* within each company division are those who have made the most money during the game, figured as the percent increase over their base salaries during the time of the game. The objective of having both company and individual winners is to introduce conflicts between individual goals and organizational goals.

Role Descriptions

Each company is composed of at least one evaluator, two designers, two managers, two builders, and three communicators. The number of players for each role can be substantially increased as long as the proportion of players in each role is maintained. In addition, there are several Game Overall Directors (G.O.D.'s), who monitor the game, clarify rules, carry out tests, and determine winners. For more detailed descriptions of these roles, see Figure 1. The range of roles provides the opportunity for vertical conflict, and the number of players in each role (given here as the minimum) provides for the possibility of horizontal conflict. Groups are instructed to operate according to their own schemes so that potential conflicts between groups and with communications can occur. This freedom forces students to determine what information they need for their group tasks, and how and whom to ask

Designers (2)   Salary: $25,000

Designers design the product and, if necessary, redesign it on the basis of tested performance, management directions, etc. Precisely how designers operate in the company is their decision. They are subject to the general rules and to additional rules that managers may receive at the start of the game.

Managers (2)   Salary: $30,000

Managers are middle management. Their function is to facilitate the designing, testing, communicating, and costing of the assigned task. They keep records of costs and are responsible for preparing and submitting the final bid. Managers are subject to the general rules as well as to any additional rules received at the start of the game.

Builders (2)   Salary: $15,000

Builders build the product the designers have designed and modify it on the basis of changes ordered by the designers throughout the game. Precisely how builders operate is their decision. They are subject to the general rules and to additional rules that managers may receive at the start of the game.

Communicators (3)   Salary: $20,000

Communicators (1) carry messages between managers, designers, and builders and between managers and G.O.D.'s; (2) purchase and transport material; (3) transport the product to and from testing; and (4) observe and report testing. Communicators are subject to the general rules and to additional rules that managers may receive at the start of the game.

Evaluators (1)   Salary: $50,000

Evaluators represent top management. They have two tasks. They initiate the playing of the game; that is, they are responsible for getting the operation started. In addition, they evaluate the performance of all individuals in their respective companies twice during the game by assigning bonuses of 0, 5, 10, 15, 20, or 25% of the individual's base salary. The evaluator of the winning team receives a 75% bonus.

Note: G.O.D.'s are in charge of the game. They sell material, keep cost records, judge that tests are properly carried out, and determine winners.

Figure 1. *BIAS* role descriptions.

for it. And asking for information can become expensive both in time and money. The only constraints on the actions of all players are covered in the general rules.

General Rules

The general rules that follow are held to a minimum to make it easier to keep them in mind and to increase the uncertainty of the game.

1. Managers, designers, and builders may not leave their rooms; communicators and evaluators may move between rooms.

2. Communications between managers, designers, and builders and between managers and G.O.D.'s are handled by communicators.

3. Managers alone may request clarification of game rules and procedures, and only from a G.O.D. Requests must be made on a standard consulting request form. The first two requests are considered in-house R&D and are free. Additional consultations are billed at $20,000 per minute.

4. Evaluators must turn in first bonus sheets no later than 35 minutes after the game has begun. Second bonus sheets may be submitted at any time after 40 minutes, but no later than the time at which the bid sheet is submitted.

5. Some forms (for example, the bonus and salary record sheet) serve multiple purposes. Multiple-use forms contain the following symbols in the upper right-hand corner: C/C (company copy), A/C (auditor copy), E/C (evaluator copy). The recipient of a given form is indicated by circling the appropriate symbol. In addition, the bonus and salary record sheets contain the words "First Bonus" and "Second Bonus" to be circled as appropriate.

Software and Hardware

The software consists of the eighteen sheets of forms, sketches, and instructions listed below.

1. Playing the Game
2. General Game Description
3. Role Descriptions
4. Role Preference Form
5. General Rules
6. Bonus and Salary Record Sheet
7. Task Description
8. Physical Layout
9. Test Description
10. Parts and Costs Sheet
11. Parts Cost Record Sheet
12. Test Cost Record Sheet
13. Consulting Record Sheet
14. Cost-per-Unit Sheet
15. Bid Sheet
16. G.O.D. Final Audit
17. Penalty Calculations
18. Individual Winner Calculations

To create a need among students for information and communication, not every player gets every sheet. Lack of space prevents
supplying examples of these items here; however, a complete set
of software and instructions may be obtained at cost from the
author.

While the hardware varies with the task, players need paper,
pencils, ballpoint pens, pocket calculators, testing equipment,
and badges to indicate both company and role. The number of
forms can be varied, but relatively numerous forms ensure that
keeping track of what is happening and making sure that all
affected personnel in the company know what is happening are
major problems. The point here is to demonstrate the need for
constant communication in the workplace.

The hardware is dependent on the design task, but it and
the task should be kept fairly simple, not only for reasons of
cost and time, but also to keep the focus on communication
problems. Experience indicates that if the design problem is
complicated, the engineering instincts of the players are aroused
and the simulation forgotten.

Beginning the Game

Players and directors meet together, introduce each other, and
begin the following sequence of events. It should be noted that
students cannot and should not be forced to play the game.
Those who are quite certain that they do not want to play can
be assigned related tasks—selling parts, running tests, keeping
the G.O.D. auditor's books, checking for violations, or even just
watching.

1. Introduction. The Game Overall Director explains the general
   purpose of the game: to experience what it is like to try to
   succeed within a complex organization when faced with
   various conflicts and communications problems.
2. Beginning: Phase I
   a. All players receive a copy of General Game Description.
   b. The players for each company are selected by the G.O.D.
      The members of each company then assemble together.
   c. All members of each company receive a copy of Role
      Descriptions. Members rank the roles they would like to

play, in descending order of interest, on the Role Preference Form and hand this form to the G.O.D.
   d. The G.O.D. assigns roles within each company, based as far
      as possible on player preferences. Role, company, and
      names are recorded on a company organization form and
      coded identification badges are issued.
   e. Managers, communicators, and evaluators go to the man-
      agers' room, designers to the designers' room, and builders
      to the builders' room.
   f. Evaluators, managers, communicators, designers, and
      builders are given their packets, which they are free to
      open and examine.
3. Beginning: Phase II
   Evaluators initiate the action and the game begins.
4. Once the game has begun, it operates under the general rules.

After Phase II, no two versions of the game are the same. If you
are unfamiliar with simulation games and can find no colleague
who is familiar with them to help you, run at least two trials.
Do the first with only one team because it is cheaper. Special
problems can be dealt with on the spot (or noted to be brought
up during debriefing). If things get confused, the game can be
ended and discussed. Restrict the second run to two teams, and
again it may not go particularly smoothly. Once through it,
however, you should have enough feel for the game to be able
to run it with confidence.

A Sample Task

Assume that your company is bidding competitively to obtain a
lucrative government contract to build 200,000 units of a vehicle
with specified performance characteristics. In order to submit a
bid, you need to design, build, and test a prototype vehicle according to the following specifications.
   Design and build, from the available material, a vehicle that
will carry a specified object down a specified test track so that at
the end of the test run the vehicle and its load are intact. A Parts
and Costs Sheet of the available material is attached. A one-pint
mason jar filled with twelve ounces of water is the specified
object. See Parts and Cost Sheet for costs. Side elevations and
top views of Test Track 1 and Test Track 2 are attached. The

costs for each test track run are not given on the Parts and Cost Sheet. Finally, a vehicle is "intact" when, at the end of Test Track 1, it and its specified object travel at least three feet down Test Track 2.

The winning bid will be determined by the G.O.D. auditor, who will verify all information on the Bid Sheet. Only *one* Bid Sheet per company will be accepted.

While this may seem a simple task, the vagaries of Tinker Toy quality control are such that a vehicle constructed from those materials has a low probability of surviving a run on even a very simple test track. Whatever task you or your students invent, *try it out* ahead of time—some tasks that sound simple turn out to be incredibly difficult, and vice versa. Again, make sure that the task is simple enough so that players are not lured away from the communications problems.

Conclusion

As described, *BIAS* takes about two and one-half hours to introduce, play, and debrief and requires a minimum of four rooms. With modifications it can be introduced in about twenty minutes of one class period, played in fifty minutes of the next, and debriefed in the third. With suitable changes in the rules, it can be played in one large room.

No matter how it is played, *BIAS* is a simulation game designed to begin the process of making students in technical communication courses aware of the organizational context in which most of them will be working and of the influence this context has on communication problems. Once aware of these problems, students are ready to meet challenges that the organizational writing tasks in technical communication courses offer.

Notes

1. For a more detailed explanation of the need for and the problems with simulations in technical writing courses, see my "Writing as Engineers and Writing in Class: Simulation as Solution and Problem," in *Technical and Professional Communication: Teaching in the Two-Year College, Four-Year College, Professional School*, ed. Thomas M. Sawyer (Ann Arbor, Mich.: Professional Communication Press, 1977), pp. 155–66.

2. *BIAS* was developed with the help of Russell Stambaugh of The University of Michigan Extension Gaming Service. Funding for this help and for game materials was provided by a Faculty Fellowship Award from The University of Michigan Center for Research on Learning and Teaching.

3. See David R. Krathwohl, Benjamin S. Bloom, and Bertram B. Masia, *Taxonomy of Educational Objectives. Handbook II: Affective Domain* (New York: David McKay, 1964).

19 Writing for John Q. Public: The Challenge in Environmental Writing

Gretchen H. Schoff
University of Wisconsin—Madison

Students planning careers in environmental fields need highly developed communication skills in order to interact effectively with their audiences—audiences always in conflict and frequently hostile to technology. An assignment designed to develop the "translation" skills these students will need is provided. For purposes of discussion, a report prepared for the Citizens Advisory Councils by a member of the Wisconsin Department of Natural Resources is included.

Students who choose environmental engineering careers with such organizations as departments of natural resources, conservation agencies, or legal and paralegal arms of government regulation soon find themselves negotiating the troubled waters of bureaucracy and public concern. Environmental decision-making inevitably relies on information from the scientific community, but that information must often be "translated" for many different kinds of public audiences and for use in the formation of public policy.

Engineers who come straight from industry or from the university to a governmental department may be quite unprepared for such intricacies as shared public financing, lobbying, or political power struggles. They may even find themselves longing for the relative simplicity of industry. They quickly realize that it is one thing to provide engineering expertise for a private company that sees an engineering task only as embedded in the business direction of the company, its products and its profits. It is quite another matter to deal with the jungle of funding—federal, state, or county—as embedded in the goals of society at large, and dependent upon laws, regulations, and decisions of legislators who in turn must read the minds and hearts of their constituents.

And it is in this second instance, the matter of public input and concern, that the environmental engineer faces a most difficult task. However idealistic the motives for choosing environmental engineering, the engineer often must deal with a general atmosphere of public mistrust. "Engineers are the waste makers, the designers of industrial plants that pollute air and water, the builders of highways that scar the landscape," says John Q. Public. And engineers must therefore prove their respectability—most often by the responsible communication of technological information. Moreover, many federally-funded environmental projects require proof of public participation in decision-making through advisory councils and public hearings. This requirement forces engineers into face-to-face situations where they must explain in a positive way what the engineering community can do to maintain or improve the environment; further, they must often explain an engineering task of considerable complexity to a lay audience with no engineering expertise. To make the task more difficult still, these engineers must prepare oral and written presentations in cooperation with engineers of other discplines, as well as with administrators and public servants who deal principally with legal, financial, or jurisdictional aspects of environmental problems.

What are the facts for a writer faced with the challenge of writing for John Q. Public? First, potential audiences are extremely diverse, ranging from a tough, hostile audience of strangers to a kindly disposed but technically naive group. Second, the writer is faced with making technical material understandable, choosing what to include, what to leave out, and then finding the proper levels and modes of language to preserve essential information.

The temptation for both the writing teacher and the student is to approach "translation" writing in scattershot fashion. The instructor may be tempted to rely on a hasty one-liner such as, "Rewrite a description of a technical task for a general audience." The student will probably respond in kind because he or she interprets the assignment to mean "leave out technical jargon" or "simplify."

The rationale for the writing assignment described below grows out of two assumptions: (1) that systematic, in-depth analysis of potential audiences needs to be a conscious and continuing process for technical writers, especially when they are

writing for nontechnical people; and (2) that one of the magic
keys of technical writers is analogy, especially the selection
of appropriate metaphors for and comparisons between complex
technical tasks or equipment and everyday occurrences or familiar
objects.

These two assumptions are built into the writing assignment.
Students are presented with a clear purpose statement and a
description of the rhetorical situation and audience for the report.
They are given steps to follow in preparing the report, steps
accompanied by questions to encourage them to think analogically.
Finally, the evaluation procedure systematically checks for the
presence of the elements described above.

Chemical, civil, industrial, electrical/and computer science,
nuclear, mechanical, remote sensing (aerial and satellite surveying)
—in brief, almost all the traditional engineering fields now offer
job opportunities in the environmental field. It is for these engineering students, both undergraduates and graduates, and for
non-engineering technical writers in environmental professions
that this assignment is designed.

The Assignment

Many writing teachers fear that too many instructions damage
creativity and spontaneity for the writer. My own experience has
been that students appreciate clearly typed, fully defined instructions and preliminary class discussion on the interpretation of
those instructions. I give students a typed copy of the assignment
at least a week or two in advance of its due date and include the
due date on the top of the instruction sheet. A typical assignment
is given below.

Due Date: _____

1. *Purpose.* The purpose of this assignment is to test your ability
   to "translate" a complex technological task into language
   comprehensible to the lay audience defined below and to so
   describe the task that the audience understands its purpose,
   its relevance to environmental preservation, and its desirability
   as public policy.
2. *The rhetorical situation.* You are an engineer employed by a
   Department of Natural Resources involved in water quality
   planning for an entire state. To receive federal funding, such
   planning requires proof of citizen participation in the form of
   Citizens Advisory Councils. These councils are typically

composed of city council members, mayors, and city and county officials. Their chief concerns are local, municipal, or county. The education of their members may range from eighth grade (or below) to college or advanced degrees. They are concerned about effects on local quality of life: environment, jobs, taxpayer costs. You will have little time or available information to do an advance or in-depth audience analysis of these councils, other than to know that, in general, they have scant understanding, if any, of the technical questions involved in water quality planning.

3. *Steps in preparing the report*
   a. Prepare a 4–6 page report to be sent to the Citizens Advisory Council, explaining an engineering task that helps to assess, improve, or monitor water quality. Here are some possibilities: a test for dissolved oxygen downstream of a filtration plant, a measurement device for chemical pollutants, the use of mathematical computer models for river profiles, the aerial surveying and monitoring of rivers or lakes, the design features of a filtration plant that will require partial municipal funding.
   b. As you prepare this report, ask yourself the following questions: Have I explained *what* this does, *why* it is done, *how* it is measured? Have I used analogy to translate engineering language (especially the language of chemistry, computers, or mathematics) to the language of a lay audience? Have I shown that the engineering task is an important component of maintaining water quality? Are there graphics or diagrams that would make the presentation clearer?

Preliminary Discussion of the Assignment

Copies of the assignment in hand, we go through its three parts to amplify what is intended, relating the discussion whenever possible to particular chapters or sections of the textbook. Here is a synopsis of the points I try to cover.

1. *Purpose.* I re-emphasize that this report is not from one technician to another (such as a chemical engineer to a civil engineer), nor is it from one level of a complex organization (such as its technical staff) to another (such as its managerial or administrative staff). Instead, students are being asked to explain a technical task or piece of equipment in terms that a nontechnical person can understand. They should not assume, for example, that readers will know why a dissolved oxygen test is useful. They must, instead, explain why it is useful. If the class has students from several engineering

disciplines, they can try out potential topics on one another, attempting to discover what might prove difficult for a lay audience to understand.

2. *The rhetorical situation.* Few students understand without help the subtle relationship between writer, report, and audience. The difficulty in this assignment is that the writer cannot know the audience in detail or with certainty ahead of time. Frequently, environmental reports end up in the hands of readers who are unaware of the context and implications of the report. What the writer can do ahead of time is to assume a range of educational levels within the audience and, therefore, a broad spectrum of concerns.

It is not enough to tell students that the audience will be a mixed bag. They need to get a feel for what that term means. One way is to ask the class to list the probable types of citizens on advisory councils. These may be political appointees whose foremost accountability is to their constituents (aldermen, county board officials, mayors); environmentalists interested in preserving wetlands, fishing waters, or shorelines (individuals or representatives of environmental groups); city and county employees (sewage treatment plant supervisors, county treasurers, game wardens); civic-minded citizens who fear loss of local autonomy or excessive regulation.

When such a list is on the chalkboard, students get a much better understanding of the range of interests. This diverse group must be "in their minds" as receivers of the written material they are about to produce. Some students find it helpful to imagine themselves talking to someone they know from their hometowns who might be a likely member of such an advisory council.

3. *Steps in preparing the report.* The actual writing of the report can be made easier if students ask themselves certain questions. The first question: "What shall I write about?" The assignment suggests topics for chemical engineers (dissolved oxygen or pollutant tests), for mechanical engineers (measurement devices), for computer engineers (computer modeling), for civil engineers (design features), for environmental engineers (aerial or satellite surveying). And this list is not prescriptive.

The second question students ask is "How do I translate 'down' to a nontechnical level?" A classroom exercise in how to draw analogies between technical and nontechnical material is helpful. (See, for example, the analogy drawn between a sentence and an equation in the report that follows, "Everything You Ever Wanted to Know about River Models . . . and Then Some." Analogies come easily for some students, but for others, practice is necessary. Household objects and natural processes are good sources for analogy.

Evaluation

Evaluation procedures depend on teacher preference. Some instructors rely on individual conferences almost exclusively, some annotate papers directly on the copy submitted by the student, and some use a separate evaluation sheet. For this assignment, I have found a combination of annotation procedures to be useful. Using the evaluation questions listed below as a guide, I read the paper, checking for the presence of key elements. Students who effectively manage these elements have, by definition, paid attention to the nature of the audience.

1. Does the report state *what* the procedures, tasks, or equipment are, *why* they are used, *how* they are measured?
2. Does the report clearly translate mathematical, computer, or chemical language through appropriate analogies, metaphors, or visuals?
3. Does the report explain the relationship of the engineering task to the larger question of water quality (or to other related systems such as aquatic ecosystems)?

If a student feels that something which I have judged to be obscure or unnecessarily complex is actually clear, I test the material with a student from another discipline, a willing colleague, or someone untrained in the writer's discipline who will take a few minutes to read the piece and react. If you try this test, ask your reader these questions:

1. Could he or she tell you, in lay terms, the *what*, *why*, and *how* of procedures and equipment?
2. Where did he or she founder? Where did confusion occur?
3. Was he or she convinced of the necessity and desirability of the procedure? If not, why not?

The results of this exercise are usually revealing to the student. If he or she must jump in to amplify orally what is written, then your judgment of the obscurity or undue complexity of the piece is likely to have been fair.

Corollary to the Writing Assignment

The State of Wisconsin Department of Natural Resources is currently involved in water quality planning for the entire Wisconsin River Basin, and Citizens Advisory Councils much like those described in the writing assignment are participating in the planning. The report that follows was prepared for distribution to those councils by a member of the Water Quality Division of the Department of Natural Resources.[1] Its author was frank in admitting that the preparation of the report had been extremely difficult and had sparked considerable discussion among engineers.

Students might begin by asking the evaluation questions given earlier. The report may also be used as the basis for an extended discussion of the very questions that troubled the writer of the report as they do students: Are the analogies apt and effective? If not, what would you do instead? Is the presentation too simple? Too involved? (Perhaps the title reveals the technical writer's own ambivalence: "Everything You Ever Wanted to Know about River Models . . . and Then Some." Is there too much "and then some"?)

Everything You Ever Wanted to Know
about River Models . . . and Then Some

There are many ways of representing the real world. Photographs, sketches, sculptures, and paintings are just a few of the ways we can produce visual images of objects. Each image must contain enough information about the object so that the viewer can recognize it. The image is a model, a representation of some real-life object.

Another way to represent the real world is through words. Here the information is not as direct as a visual image. The reader must take the written symbols and translate them into an image in his or her mind. The written description can also be a model.

Mathematical models are a third way to represent objects. Through a mathematical model it is possible, for example, to describe the way the planets move around the sun and then to predict where each planet will be at a specific time.

In a similar way, a mathematical model of a river is a tool that represents the relationships between the biological, chemical, and physical activities occurring in the river. Today, a model is the major tool for water quality planning. The use of a model makes it possible to predict how the river will respond if a proposed action is taken. In this way, proposals can be "tested" before they are actually put into effect.

The first attempt to develop a model of a river occurred in the early 1920s, before computers were available. This first model was a simple statement, in mathematical form, to predict how much oxygen would be consumed by wastes added to a stream. This equation, known as the Streeter-Phelps equation, is still used. Other factors, however, also affect the river's water quality, and the development of high-speed computers has made it possible for today's models to include many more of these relationships. Many people, working around the country, have contributed to the vast improvement in computer modelling over the last decade.

In addition to a large computer system, a river model needs two things: (1) mathematical equations to describe relationships within the river and (2) observations of the "real life" river being modelled.

Mathematical Equations: Sentences in a Different Language

The language of a river model is mathematics and, like all "foreign" languages, math takes work to be able to understand it. We will attempt to demonstrate how a model works using familiar illustrations.

Equations are the sentences of mathematics. Equations are also something each of us uses everyday, perhaps without knowing it. We use an equation, for example, when we calculate how long it will take to drive from Rhinelander to Madison (Time = Distance $\div$ Speed). Or, at the risk of simplifying things too far, consider how you might predict your grocery bill if all you bought were apples and pears. This equation would help you: Total Cost = (price of one pear) $\times$ (number of pears you buy) + (price of one apple) $\times$ (number of apples you buy). This equation contains a set of instructions that tells you the steps you need to go through to find the cost of your purchases. It can also tell you how the total cost will be related to the number of apples and pears you buy.

There are two nice things about equations. One is that they are a shorthand way of describing a complicated situation. Our equation above, however, is still quite long. It would be convenient to make it a bit shorter. If we define the terms this way, $\$$ = total cost, C_p = cost of one pear, P = number of pears you buy, C_a = cost of one apple, and A = number of apples you buy, we can shorten the equation to $\$ = C_p \times P + C_a \times A$. Once we have defined what each of the letters or symbols means, we no longer have to write the whole thing out.

The second nice thing about an equation is that it can be used in many situations. The equation $\$ = C_p \times P + C_a \times A$ works, no matter what the cost of pears or apples or how many you buy. If pears cost five cents each and apples cost ten cents each, the appropriate prices are "plugged in" and the equation becomes $ = 5¢ × P + 10¢ × A.

The 5¢ and 10¢ are called *coefficients*. In this case, we could go to the grocery store and look at the price tag to determine the "correct" value. More important, they are numbers that are independent of the number of apples and pears you decide to buy—they are set by the grocer. They are the key to describing how total cost is *related to* the numbers of apples and pears. This tells you that your total cost will go up 5¢ for each pear you buy.

On the other hand, number of apples and number of pears are called *variables* because they can change, or vary, depending on your decision. Thus, if you decide to buy 3 pears and 2 apples, the following equation results:

$ = 5¢ × 3 + 10¢ × 2
$ = 15¢ + 20¢
$ = 35¢

Of, if you decide to buy 4 pears and 1 apple, the equation is as follows:

$ = 5¢ × 4 + 10¢ × 1
$ = 20¢ + 10¢
$ = 30¢

The answer depends on which variables (how many apples and pears) you choose. The important thing is that the same equation can be used in many situations.

We have just built a simple model of your grocery store bill. The model could be expanded to include every item in the store.

From Apples and Pears to River Models

What has all this got to do with a river model? A river model is also based on coefficients, variables, and equations. But instead of one equation, there are many. In addition to simple instructions such as "multiply" and "add," a river model uses complex instructions that require advanced mathematics. Instead of apples and pears, some of the variables are the amount of water flowing in the river, the water temperature, and the amount of wastes going into the water. When all of the equations are calculated, the model *predicts* what the dissolved oxygen of the river will be at each place along the river. It allows you to test the reactions of the river to many different conditions of flow, temperature, and waste loads.

One problem in designing a model is identifying the important relationships which make up a river system. The studies of biologists, chemists, and engineers are helpful in this effort.

After the relationships are identified, it remains a problem to determine the accurate values for the coefficients (the cost of an apple or a pear in our first example). For an apple or a pear we can go to the store and find out the price, but it is a much harder task, for example, to find out how fast oxygen can transfer from the air into the water. Different ways of measurement may be possible, and the answers may not always agree. It is often the case that you cannot say that one way is right and the others are wrong. It then becomes a matter of judgment to decide which value to use.

Calibration and Verification: The Proof of a River Model

Because different values are possible for important coefficients, it is necessary to fit the model to the real-life river. This fitting process, or *calibration*, begins with gathering information by taking detailed measurements of the river. A survey team will record the temperature, flow, dissolved oxygen, waste loadings, and many other variables up and down the river. Several surveys must be taken at different times and under varying conditions. The river modeller then tries different coefficients until the model accurately reproduces the observed data from the surveys. Calibration is the process of tuning the model, much like tuning a car until it runs smoothly.

When the model is calibrated using several data sets, it undergoes the final test of *verification*. For this, an entirely new set of data is used, a set that was not used during calibration. If the model predicts the same dissolved oxygen that the new data set measures, it is *verified*. If it cannot accurately reproduce the data, the modeller must locate the cause of the problem and correct it. After correcting the problem and recalibrating, another verification run is made. This process continues until the model accurately describes the river system.

What the Model Can't Do

A model is a tool. Like other tools, it performs some jobs very well and is entirely unsuited to others. It it as important to know what a model can't do as to know what it can do.

1. The model *cannot* tell you how much waste each discharger should be allowed. The model *can* tell you approximately how the river will react if a certain level of waste is discharged at a given time and place and under a specific set of river conditions.

2. The model *cannot* tell you how much dissolved oxygen the river should have. The model *can* let you test different ways to achieve a target level of dissolved oxygen, but the target level must be defined by a policy-making group.

3. A model *cannot* be expected to give an entirely accurate prediction. It *can* tell you, however, which factors have the greatest influence on water quality. For example, if you run the model many times and each time you keep it exactly the same except for a small change in water temperature, the results will show how sensitive the river is to changes in water temperature. This procedure is called sensitivity analysis, and it is a very important use for the river model. Knowledge of the sensitivity of the river suggests which alternatives for improving water quality are worth pursuing and which alternatives would have very few desired results.

4. Most importantly, the model *cannot* make policy. It does not tell you what is "good" and what is "bad." It cannot tell you what "should" be done. The model *does* allow you to test different policies. It can, for example, predict the water quality you might expect if the policy "there will be no point source discharges to the river" were chosen.

Similarly, it can predict how the river will react if all discharges were cut in half, if you take out a dam, if you move the discharge point of a waste treatment plant, if you increase the flow of water by releasing water stored in a reservoir, or if you add extra oxygen to the water. Each of these might become a policy, but people still need to decide if they want that policy. The model helps to determine the effect of each policy on water quality. This is why a model is essential for water quality planning. It helps to guide decisions. It cannot make decisions. A model answers the question "What if?" It does not answer the question "What should be?"

Notes

1. Permission granted by the author of this report, Monica Jaehnig, Wisconsin Department of Natural Resources, is gratefully acknowledged. She offered the report as a typical response to the challenges and difficulties inherent in environmental writing for public audiences rather than as an ideal solution.

20 Decision-Making and Problem-Solving: An Holistic Writing Assignment

Susan B. Dunkle and David M. Pahnos
Carnegie-Mellon University

The holistic problem-solving procedure suggested here shows students that the writing process is a thought process. The authors see the assignment as especially valuable for technical problems whose solutions are linked with cultural values.

In the eighteenth century, philosophy experienced a distinct split between empiricism and rationalism. What had been for Aristotle two inseparable methods of inquiry became the cores of thought for two disparate schools of philosophy. One legacy of this split remains: many technically-oriented students lack the conceptual ability to handle problems that require both a technical and a cultural solution.

In the last few decades, much effort has been devoted in American universities to teaching methods of empirical problem solving, especially in the field of engineering. While an empirical approach yields technical solutions, it cannot fully resolve those problems that are largely mediated by personal or cultural values: questions concerning the *quality* of life cannot be answered in *quantitative* terms. Educators have too often assumed that technical students would acquire in humanities courses the conceptual abilities that, when integrated with empirical problem-solving techniques, would allow them to think in holistic ways. Unfortunately, many technical students do not acquire a balanced approach to holistic thought in college. Further, the need for engineers to reason holistically may also be linked to the need for a humanizing technology, i.e., one that is responsive to social problems, desires, and aspirations.

Nowhere is this debility more evident than in writing, since written work is a clear representation of the thought process. Cultural historian Jacques Barzun has characterized the situation:

205

"We have ceased to think with words. We have stopped teaching our children that truth cannot be told apart from the right words." Using the right words implies that the student has gone through a holistic thought process.

It is not easy for teachers of technical writing to convince students that the quality of their thought depends on their ability to communicate—and that the writing process is a thought process. Students invariably argue that their written works are logical and that the instructor is either dense or overly critical.

Part of the problem is that students too often confuse problem-solving processes with decision-making processes, treating conclusions as if they were quantitative products, when they may really represent qualitative choices based on internal assumptions. Students quickly learn not to write "I think" or "I feel" in a technical report, yet they continue, as they must, to make value judgments. However, they generally have not mastered a heuristic procedure that will allow them to handle value hierarchies in a decision-making process and may not be able to separate what they can prove from what they feel. The results are marked by confused syntax, poor organization, and a conspicuous lack of transitions.

It is important, then, for students to be able to differentiate between a quantitative product and a rational choice; if they cannot, their writing will be at best unclear and at worse deceptive, for to confuse the two processes results in a genuine loss of objectivity.

In order to help students overcome this problem, we have developed an assignment format that takes them through a sequential, holistic, problem-solving process. The heuristic procedure is based on decision-making theory. We give students a systematic way to establish a value hierarchy that allows them to separate and then integrate social and technical problems. With appropriate variations in technical material, it can be used in any engineering class. (The sample assignment presented later is intended for use with junior mechanical engineers.) The format has seven steps: (1) define a goal, (2) establish objectives, (3) classify objectives according to importance, (4) evaluate alternatives in relation to objectives, (5) choose the best alternative as a tentative solution, (6) assess the adverse consequences of the tentative solution, (7) make a decision.

In general, the format forces students to explore a problem, formulate objectives, test alternative solutions, and make a decision. They must then assume a writer's role and reorganize their information to make it readily understandable to a specific audience. If students can successfully work through the holistic, problem-solving process, then it is far more likely that they will be able, in the written report, to demonstrate clear, logical organization, to make appropriate transitions that allow the reader to follow their thought process, and to distinguish clearly between quantitative findings and qualitative decisions.

The Assignment

Assume that you are working for the government and must decide whether electric or internal-combustion engine vehicles should be used exclusively in the United States. Obviously, you want an efficient (from a fuel standpoint), pollution-free vehicle using a fuel that will be available for the next hundred years. The engine and fuel should be inexpensive so that American families can continue to own at least two cars. Compare the alternatives and present your decision.

Defining the goal. When confronted with a complex problem, such as the one in the assignment, students must first get enough control over the problem to begin a systematic investigation. To guide this investigation, they must define a goal, that is, determine what they want to accomplish. Since we are concerned with holistic problem-solving, the goal statement should include expectations for both the empirical and the rational solution. For example, a partial goal for the assignment must include that the vehicle use available fuel (technical goal) and that it be inexpensive enough so that families can own two cars (social goal).

Establishing objectives. The next step in the process is to define objectives for the investigation. These objectives are the standards by which students appraise the alternatives in the assignment to make a decision. As with all standards, an objective is easiest to attain if it can be quantified. For example, a technical objective might state that the vehicle be light-weight, a statement that can be quantified by defining light-weight as between 1800–2200 pounds. A social objective might state that the vehicle be safe. At this point, students begin to see the relationship between the social and the technical. For example, we might mean by "safe" that the vehicle can withstand a rear-end collision without whiplash injury to passengers or without the gas tank exploding. Thus a social objective, in this case, can only be achieved by solving a technical problem.

Classifying objectives according to importance. All the objectives a student defines exert some degree of influence on his or her final decision. Some, however, will be of absolute and overriding importance; some will be important but not mandatory; some will be useful as a bonus but won't affect the situation a great deal one way or the other.

Objectives, then, can be divided into two categories: negotiable demands and nonnegotiable demands. Nonnegotiable demands are absolute constraints; they set standards that cannot be violated by any potential solution to the problem. For example, nonnegotiable demands might be that the vehicle use fuel available for the next hundred years, be inexpensive (under $5,000), and carry four passengers. (Note that nonnegotiable demands include both technical and social considerations.) Negotiable demands are relative constraints; they can be deleted. Some negotiable objectives will be more critical than others. To establish the importance of negotiable demands, students must weigh each one carefully. To do this, they assign a relative importance to each demand. For example, they might assign a number from 1 to 5 for each demand; 5 being an important demand, 1 being a mildly desired one. Low ranked demands, obviously, can easily be given up.

At this point, students establish a value hierarchy. They must decide whether safety is more important than cost, fuel economy more important than auto weight, passenger comfort more important than minimum external dimensions, fuel efficiency more important than rapid acceleration.

Evaluating alternatives against objectives. Students now use self-designed standards to see how alternatives stack up. First, they consider the nonnegotiable demands. If any alternative fails the test, it is out. It is as simple as that. Negotiable demands are more difficult to reconcile. To each alternative, students assign a relative weight. They can then add up the weights an alternative has collected and rank the various alternatives. The alternative with the highest total is ranked highest.

Choosing the best alternative as a tentative solution. The alternative that receives the highest weighted score on performance against the objectives is presumably the best course of action. On the basis of the evidence considered, that alternative has checked out best. It may not be a perfect choice; it may be only the best of the alternatives available.

Assessing the adverse consequences of the tentative decision. Finally, we ask students to answer this question: "If I choose this alternative, what could go wrong?" Assessing the consequences of a decision is perhaps the most important step in the process. Again, students look at both empirical and rational effects. They must determine what could go wrong from a technical standpoint, but they must also determine what the social effects of the choice might be—on the environment, on the economy, on public safety. If several things could go wrong with an alternative, and the probability is high and the consequences major, students will probably eliminate the alternative—even if it was the prime candidate. We are trying to enable students to see that to solve a problem rationally includes visualizing the impact of a decision. In some cases, technically sound ideas will be rejected for social reasons.

Making a decision based on the analysis. The last step of the process is to make a decision based on the analysis.

Writing the Report

To this point, the student has been primarily a problem solver: exploring a problem, formulating objectives, testing alternative solutions against objectives, making a decision. Now the student becomes a report writer. His or her purpose is to define the problem for readers, to recommend a solution, and to defend that solution against alternatives.

We remind our students that their readers are not essentially interested in *how* a writer went about reaching a solution; rather, readers want to know, from the first paragraph, what problem a writer intends to solve and how he or she intends to solve it.

Although the reader will not see the writer's investigative process, the organization of the report should clearly indicate that the writer arrived at conclusions through a logical, holistic sequence. The proof of this, of course, is the writer's ability to distinguish ideas clearly, to show how ideas connect logically, and to demonstrate, in writing, a sequence of reasoning that brings the reader to the writer's conclusion.

It is our experience that this problem-solving approach to an assignment not only helps students to reason in more holistic ways but also gives them a better understanding of the relationship between thinking and writing.

21 Teaching the Use of English Articles to Nonnative Speakers in Technical Writing Classes

Thomas N. Huckin
The University of Michigan

Nonnative students in technical writing courses have difficulty using English articles, and improper usage seriously affects the readability of what these students write. The author presents a step-by-step method of instruction with the following features: it utilizes an algorithm in the form of a flow diagram as both a teaching tool and a reference sheet; it covers in distilled fashion a broad range of conditions governing the use of articles; and it takes no more than sixty or seventy minutes of the instructor's time.

With few exceptions, technical writing teachers in America are sooner or later faced with the challenge of teaching technical writing to the nonnative speaker. American science and technology continue to have great importance around the world. Other countries, particularly those of the Third World, are putting increasing emphasis on the study of English as a second language. More and more students are coming to the United States to further their educations, with most of them electing a scientific or technical program of study. Many, especially those at the undergraduate level, enroll at some point in a technical writing course.

To a certain extent, teaching technical writing to the nonnative student is no different from teaching it to the American student: both have more or less the same need, and the same ability, to learn to analyze audiences, to write purpose statements, to write abstracts, to design effective technical discussions, to devise serviceable formats, to use visual aids—in short, to do those report-writing activities that are "macroscopic" in nature.

On the other hand, numerous "microscopic" activities are apt to cause special trouble for the nonnative student. I have in mind such editing tasks as choosing correct terminology, punctuating properly, using passive verbs appropriately, creating parallelism, and avoiding ambiguity. Although such matters are sometimes considered "low level," they can have as significant an impact on a report's readability as "higher level," macroscopic matters. And, needless to say, they require a high degree of proficiency in English. Since foreign students often lack that proficiency, this is an area where the technical writing teacher can obviously provide special help.

One of the most troublesome microscopic matters for the foreign student to master is the use of English articles: definite (*the*), indefinite (*a/an*), or none at all. Although descriptions of proper usage can be found, they either fail to provide adequate coverage of the many types of uses or seem so complicated that students (and teachers) throw up their hands in despair. Nowhere, to my knowledge, can one find the simplified *prescriptive* treatment that is suited to the needs and restrictions of the technical writing class.

My purpose here, then, is to describe a method that I have found successful for teaching the use of English articles in a limited time. In recognition of the fact that only a few students in a class may need this instruction, and that the teacher may therefore not want to grant class time to the subject, the method is adaptable to office-hour use, involving as little as thirty minutes of instruction.

The paper is organized as follows: (1) illustrations of the problem, (2) objective of instruction, (3) description of instruction, and (4) summary.

Illustrations of the Problem

In English, the correct choice of article is important on the grammatical level, on the semantic level, and on the rhetorical level. Failure to make the correct choice exposes the writer to doubts about his or her general competence in English—and it can also seriously affect the comprehensibility of the writing.

Despite the importance of articles, however, even the most advanced students of the language make errors with them. One student of mine, a bright and conscientious Korean doctoral candidate in Industrial and Operations Engineering, who had an otherwise excellent command of both written and spoken English, showed me a preliminary draft of his dissertation proposal with

To: Mr. N. M. Bailey, Chief Engineer
           Amoco Shipping Company
           555 Fifth Avenue
           New York, New York 10017

From: J. T. Wang, Supervisor
           Maintenance Department
           Brooklyn Ship Yard, Inc.
           P.O. Box 1001
           Brooklyn, New York 11236

Date: February 5, 1980

Subject: Repair of Defect in the shipboard Distilling Plant Griscom-
           Russel Soloshell, Low-pressure, Double (2) Effect Distilling
           Plant

Dear Mr. Bailey:

1   In your letter of February 1, 1980, you suggested that work be done on
2   the defective distilling plant in your ship, MV JEFFERSON. You expressed
3   concern about its failure to produce fresh water for boiler feed, drinking,
4   cooking, bathing and washing. You also suggested that tests be conducted
5   to determine if it needs to be replaced. This letter contains the result
6   of our investigation.

7   Our finding indicates the cause of the problem is the weight loaded
8   reducing valve, which is sticking. Correct resetting of the valve
9   followed by a short period of distillation produced the useful water.
10 But the replacement of this defective reducing valve with a new one will
11 solve the problem. Therefore it is not advisable to replace the
12 distilling plant: since there is no major damage done to it.

13 The trouble shooting in the distilling plant consisted of using indicators
14 such as thermometers, pressure gauges, gauge glasses and salinity indica-
15 tors to determine where the problem, that was occurring, originated.

16 We found the salinity indicator on the second effect tube nest indicating
17 high salinity. This led us to prime the first effect. By the checking
18 the temperature in the first effect, a high temperature reading led us to
19 check the pressure after the weight loaded reducing valve. It was
20 sticking with the unbalanced pressure of superheated 5 psi steam.

21 A reset of the valve and light production yielded a condensate not exceed-
22 ing the safe limit of 0.25 gpg.

Figure 1. Problems with articles in a student's informal report.

many article errors. When I pointed these out to him, he readily acknowledged that he had "always had trouble figuring out which article to use." To get an idea of the kinds of errors he was making, consider first these examples, where the mistake occurs on the grammatical level:

1. In frequent exertions, recovery period between exertions will become shorter, and less recovery from prior exertions would be expected.
2. Static component is an important determinant for fatigue in types of moderate or heavy dynamic work. . . .

In these cases the student failed to observe that words like *period* and *component* are "countable nouns" in English (as opposed to "noncountable nouns" like *osmosis* or *surgery*) and, as such, require an article when occurring in the singular.

On the semantic level, errors like the following could be found:

3. The type of activity to be investigated is static contraction of bicep muscle in alternating work cycle and rest cycle.

Since the meaning of *muscle* varies according to whether the word is being used as a countable noun or as a noncountable noun, the student's failure to use *a* or *the* in this case led me to conclude (incorrectly) that he was talking about bicep muscle *tissue*, not bicep muscles *per se*.

On the third level, that of rhetorical analysis, errors like the following occurred:

4. The measure of muscle fatigue is constructed by functional attributes of muscle, i.e., motor units. . . .

Here the use of the definite article implies that the reader knows, or at least can determine, what "measure of muscle fatigue" is being referred to. But I as reader found that the referent was not at all readily determinable. In order to figure out what "measure" he was talking about, I had to skim backwards through two full pages to where I located mention of "a measure of muscle fatigue." The intervening discussion had erased the mention of this concept from my memory, and my student had provided no clues by way of rhetorical patterning or reiteration to prepare me for the reappearance of the concept.

As another example of rhetorical-level error, consider the informal report submitted as an assignment in my technical writing class by a senior from Nigeria majoring in Naval Architecture and Marine Engineering (Figure 1). I had no trouble reading this report until I came upon the phrase "the useful water" (line 9). The use of the definite article led me to believe that the writer somehow expected the reader to know precisely which quantity of "useful water" he was talking about. And yet no mention of such a quantity of water had been made, nor would it seem likely that the reader would immediately know of such a referent from the situational context. Note, in particular, that the only previous mention of "water" concerns the *absence* of water; since such negative mention fails to specify a particular and identifiable quantity of water, it cannot serve as an antecedent for "useful water" in paragraph two.[1]

I decided to force an interpretation whereby the phrase "the useful water" *did* refer to the phrase "fresh water for boiler feed, drinking, cooking, bathing and washing" in paragraph one. Fortunately, this strategy worked: the student, as he informed me later, had indeed been trying to say something like, "Correct resetting of the valve followed by a short period of distillation produced *fresh water as desired.*" (Notice the lack of a definite article in this revision.)

This kind of rhetorical confusion owing to a misuse of articles is commonplace even among foreign students who have been in the United States for many years. Perhaps the most characteristic manifestation of it occurs in the overuse of the definite article. For instance, notice how many times the definite article is used in the report shown in Figure 1. This use creates the impression that there is a high degree of shared knowledge between writer and reader.[2] I doubt that the writer intends this effect. Having seen much nonnative writing, I believe instead that he is simply unaware of the nuances and implications of definite articles.

The misuse of articles significantly affects readability. At best, it gives the reader an impression either of linguistic incompetence or of simple inattention to detail on the part of the writer; at worst, it leads to complete (if momentary) confusion. In my experience, few nonnative students have the confidence of native students in their ability to use articles correctly; few would not stand to benefit from instruction in this area.

Objective of Instruction

The general objective of the instruction described here is to teach students to use articles correctly and effectively. However, at least two important contraints stand in the way of achieving this objective: first, only a very limited amount of class time can be devoted to the subject, especially if the class has a mixture of nonnative and native students; and second, since the factors affecting article-choice are many and varied, ranging from grammatical to situational (i.e., cultural) ones, no amount of drilling can teach correct usage.

In light of these constraints, the approach taken here aims at a more limited, but also more specific, objective—namely, to provide students with an algorithm (in the form of a flow diagram) so that they can determine on their own which article to use in a given instance. Further, much of the instruction can be completed in a fifty-minute period, conceivably during the instructor's office hours if necessary.

Description of Instruction

The method of instruction is divided into three phases: pretest, instruction, and evaluation. If possible, the three phases should be spread out over at least three days, even though the total amount of time is only about an hour.

Pretest (5–10 minutes)

The first phase consists of a cloze-type pretest (Figure 2). Give this pretest to students you suspect of having trouble with articles, instructing them to insert the best choice (*the, a/an,* or *no article*) in each blank. This task takes no more than five or ten minutes. For your convenience, the answers are supplied in italics on the pretest.

Grade the test before the next class period, marking each deviation. Then, by referring to the penalties column in Figure 2, subtract the appropriate number of points for each deviation. Subtract the total number of penalty points from 100. Students who score under 80 need the instruction that follows.

Instruction (50 minutes)

Make a copy of the flow diagram (Figure 3) for each student. Begin the class with a few introductory remarks about the importance of articles in English; the earlier discussion here provides some examples. Such introductory remarks are needed because students whose native language does not contain articles tend to

Pretest and Answers   Penalties

    Since ___1. *the*___ end of World War II more   1. an or Ø (−4)
than 9½ million veterans have bought ___2. Ø___   2. the or a (−4)
homes with the aid of ___3. Ø___ GI loans.   3. the (−3), a (−4)
___4. *The/A*___ great majority of these veterans   4. Ø (−4)
have bought ___5. Ø___ soundly constructed   5. the or a (−4)
homes and are now making ___6. Ø___ regular   6. the (−3), a (−4)
repayments on their mortgages as satisfied
homeowners.

    However, ___7. *a*___ relatively small per-   7. the or Ø (−4)
centage of these veterans have had just cause
to be dissatisfied with ___8. *the*___ outcome   8. an or Ø (−4)
of their venture into ___9. Ø___ homeowner-   9. a (−3), the (−4)
ship. It is hoped that this pamphlet, by
stressing ___10. *the*___ important things that   10. Ø (−1), an (−4)
___11. *a/the*___ prospective homeowner should   11. Ø (−4)
know, will help to reduce ___12. *the*___ number   12. a (−2), Ø (−4)
of such cases in ___13. *the*___ future.   13. Ø or a (−4)
    Buying ___14. *a*___ home is usually   14. the (−1), Ø (−4)
___15. *the*___ most important financial trans-   15. a (−2), an (−4)
action in ___16. *the*___ lifetime of ___17. *the/an*___   16. a (−3), Ø (−4)
average family. Before you decide to buy   17. Ø (−4)
___18. *a*___ house, therefore, you and your   18. the (−1), Ø (−4)
family should be certain that you are:
    Getting ___19. *the*___ right house—   19. a (−2), Ø (−4)
___20. *the*___ one that suits ___21. *the*___   20. Ø (−1), a (−4)
needs of your family.   21. Ø or a (−4)
    Aware of ___22. *the*___ responsibilities   22. Ø (−2), a (−4)
that ___23. Ø___ homeownership brings.   23. the (−2), a (−4)
    ___24. *The*___ main purpose of ___25. *the*___   24. A (−3), Ø (−4)
GI homeloan program is to help ___26. Ø___   25. a (−3), Ø (−4)
veterans finance the purchase of ___27. Ø___   26. the (−3), a (−4)
reasonably priced homes at ___28. *a*___   27. the (−3), a (−4)
favorable rate of interest. It encourages   28. the (−3), Ø (−4)
___29. Ø___ private lending institutions   29. the (−2), a (−4)
to make ___30. Ø___ bigger loans than they   30. the or a (−4)
otherwise could by guaranteeing part of
___31. *the*___ loan.   31. a (−1), Ø (−4)

Figure 2. Pretest, answers, and penalties. Source: Veterans Administration, *To the Home-Buying Veteran*, VA Pamphlet 26-6, revised (Washington, D.C., 1977), p. 1.

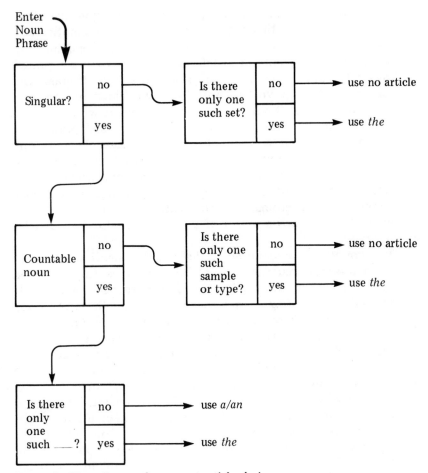

Figure 3. Flow diagram for correct article choices.

overlook their importance in English. Do not assume that all of the students who need this instruction are equally well motivated.

Next, hand out copies of the flow diagram, explaining how to use it in making correct article choices. Go over the pretest answers, checking some of them against the flow diagram; this preparation will help you to anticipate the kinds of problems students are having. Students are usually familiar with the word *noun*, but not with the term *noun phrase*. A noun phrase, for our purposes here, is a noun plus any of those elements that can modify it: adjective, quantifier, even a relative clause. Examples of noun phrases having the same head noun would include "chemical changes," "many changes," and "changes that involve the combining of atoms." Additionally, an unmodified noun acting as the subject or object of a sentence is also considered a noun phrase.

The countable/noncountable distinction mentioned earlier refers to the fact that the class of English nouns can be divided into two subclasses, those perceived as referring to denumerable (countable) entities (e.g., *pencil, page, table*) and those perceived as referring to bulk or mass (uncountable) entities (e.g., *water, sand, traffic*). This distinction has certain formal manifestations in English. For example, countable nouns are never preceded by the quantifier *much* (*much pencils, much pages*), but noncountable nouns commonly are (*much water, much traffic*); conversely, noncountable nouns are never preceded by the quantifier *many* or by a number (*many waters, twelve traffics*), while countable nouns commonly are (*many pencils, twelve pages*).[3] Countable nouns readily occur in the plural form, noncountable nouns do not (*waters, sands, traffics*).[4] Of importance to the present discussion, of course, is the fact that singular countable nouns can take the indefinite article (*a page, a table*) while singular noncountable nouns cannot (*a water, a traffic*).

Many languages of the world do not make such a formal distinction between countable and noncountable nouns. As might be expected, native speakers of such languages often have trouble determining whether a given noun in English belongs in the "count" or "noncount" category. The difficulty becomes especially acute when abstract concepts are involved: for example, *projection* is usually treated as a countable noun ("A projection was made.") whereas *dejection* usually occurs as a noncountable noun ("Dejection is one of the symptoms.").

My own strategy in teaching the count-noncount distinction is to concentrate at first on concrete, everyday nouns like *pencil*, *table*, and *water*. I point out that the countable category tends to include objects that have a definable form or shape, while the uncountable category usually embraces amorphous objects. (An amorphous object can be given form or shape by being put in a solid container, however, in which case the noun phrase as a whole becomes countable, e.g., *a bucket of water*.) Students usually have no trouble making the distinction as long as the discussion is restricted to concrete, everyday objects.[5] However, students often become confused by more abstract concepts. It is, for example, difficult for students to conceive of a countable noun like *projection* as having definable form or shape; they may argue that, for them, it is just as amorphous as *dejection*, which is uncountable. While conceding that the distinction is not always clear, I point out that *within our culture* a projection is usually thought of as a discrete process or act defined by purpose, time, and other parameters; dejection, on the other hand, is perceived as a nondiscrete state of being, not very readily definable. Although this explanation does not serve for every difficult case, it can serve as a guideline for many cases. Beyond that, I know of no general principle that would obviate the need to learn the categorization of each noun individually.

Linguists and philosophers have long noted that the function of the definite article is to signify that the referent of its associated noun phrase is, in some way, "unique." In some cases, this uniqueness is intrinsic to the noun phrase. For example, superlative adjectives (*highest, slowest, most important*) automatically mark their accompanying nouns as referring to a unique entity: only one such entity can be highest or slowest or most important in any one particular universe of discourse. Thus, it is no accident that nouns modified by superlative adjectives always take the definite article.[6]

In other cases, the uniqueness is inherent in the situational context. For example, we all know that World War II ended only once, and most of us know that there is only one GI homeloan program. Hence we say, as in Figure 2, "*the* end of World War II" and "*the* GI homeloan program"; we do not say "*an* end of World War II" or "*a* GI homeloan program."

In some cases, this sort of shared knowledge between writer and reader (or between speaker and listener) does not exist; and yet the definite article can still be used effectively to indicate uniqueness. For example, if a reader of the VA pamphlet to which we have been referring did not know there was only one GI homeloan program, he or she could still be led to that understanding, simply by the writer's use of the definite article. Indeed, it would be difficult for the reader to interpret the phrase "the GI homeloan program" in any other way. To take an even more mundane example, suppose you were told by a stranger that "the neighborhood drunk makes a racket at night." Though you may have no other knowledge of this person's neighbors, you would be perfectly justified to conclude, solely on the basis of the definite article, that there is only one drunkard in that particular neighborhood.

In yet other cases, the uniqueness of a noun phrase referent resides in its having been previously brought to the reader's attention (usually in the immediate context). This "second-mention" condition for definite article usage has much in common with the conditions for pronoun usage and is illustrated in the continuation of the passage:

> A GI loan to purchase, construct, alter, improve, refinance or repair a home cannot be approved by the Veterans Administration unless you certify that you occupy or intend to occupy the property as your home.

Here the noun phrase "the property" refers uniquely to whatever home is being referred to in the first part of the sentence; on this purely intratextual basis, the phrase can take the definite article or be replaced by a pronoun:

> A GI loan to purchase, construct, alter, improve, refinance or repair a home cannot be approved by the Veterans Administration unless you certify that you occupy or intend to occupy it as your home.

Another, final case of uniqueness concerns one type of generic noun phrase, as exemplified in a sentence like the following: "The beaver has a flat tail and builds dams." Here the writer is referring to the beaver as a species, that is, to the entire set of beavers. Just as there is only one such species, there is only one such entire set—hence the uniqueness required for the definite article.[7]

We have seen that the "uniqueness criterion" of definite article usage can be satisfied in a variety of ways. However, rather than represent each of these conditions in the flow diagram, I have simplified matters: the three final-state questions ("Is there

only one such set?" "Is there only one such sample or type?"
"Is there only one such _____?") ask essentially the same thing,
which is simply, "Does the noun phrase in question have a unique
referent?" As with many efforts to simplify, however, there
is a cost involved, and the simplification of the flow chart places
an extra burden on the instructor to explain and resolve unclear
cases.

Having distributed and discussed the flow diagram, the next
step is to put the diagram into operation. Distribute fresh copies
of the pretest and use the diagram to analyze as many of the
article choices as time permits.

To illustrate the procedure, consider the first few instances.
The first blank is associated with the noun phrase "end of World
War II." We answer "yes" to the first question on the flow chart
and move to the cell below it. Here again the answer is "yes,"
since the noun *end* is categorized in English as countable: a war
could end temporarily and then start up and then end again, thus
having two ends. Note also that yardsticks and pencils, for example, have two ends, or that a movie script might have two or
more ends (endings). When we move on the next question, however, we know that there was only one end of World War II, so we
answer "yes" to the question at the bottom and thus arrive at *the*
as the correct answer.

The second blank is associated with the noun *homes*, which
should be interpreted as referring to a random subset of the larger,
generic set. Thus, the answer to the question "Is there only one
such set?", in this case, is "no"; therefore, no article is used.[8]
Note that the indefinite article generic would also go well in this
context: "Since the end of World War II more than 9 1/2 million
veterans have bought *a home* with the aid of a GI loan."

At the end of the period, read the pretest answers that you have
not had time to analyze in class so that students can analyze them
at home. Also, return the graded pretests at this time.

Evaluation

After the instruction—preferably after several days—evaluate
student progress in two ways: (1) by devising and administering
a test similar to that in Figure 2 and using it for direct testing
and (2) by observing how students use articles in their written
assignments. The former method allows you to cover a broader
range of uses, since it confronts students with certain difficult
choices that cannot be avoided. The latter method, however,

is generally more effective as an ongoing teaching instrument since students are usually more interested in precise communication when the subject is one of their own choosing.

Summary

We began this paper by examining a number of examples of incorrect article usage by nonnative engineering students, showing how such errors seriously affect readibility. We then described a method of instruction with the following features:

1. It utilizes an algorithm in the form of a flow diagram as a tool for the teacher and a reference sheet for the student.
2. It covers, in distilled form, a broad range of conditions governing the use of articles.
3. It takes no more than sixty or seventy minutes of the instructor's time, fitting conveniently into class or office-hour scheduling.

In short, this method should prove useful to the technical writing instructor faced with nonnative speakers of English who are having trouble with English articles.

It must be borne in mind, however, that mastering the use of articles is difficult for anyone whose native language does not include them. The conditions governing correct usage in all instances are certainly more complicated and opaque than it would seem from the deliberately simplified treatment outlined here.

Notes

1. For example, imagine a conversation in which Speaker 1 says, "I really wanted to go to that concert Saturday night, but I didn't have a ticket," and Speaker 2 responds, "You mean you weren't able to go?" At this point, Speaker 1 might say, "No, I *did* go. At the last minute I found a ticket." Note the indefinite article before *ticket*. Speaker 1 would *not* say: "No, I *did* go. At the last minute I found the ticket."

2. In order for this report to be readily understood, the reader would have to be familiar with "the weight loaded reducing valve" (lines 7-8), "the second effect tube nest" (line 16), "the first effect" (line 17), "the unbalanced pressure of superheated 5 psi steam" (line 20), and so on.

3. Nouns that normally occur with noncountable interpretations can sometimes be used in special contexts where a countable interpretation is intended. For example, *many sands* might appropriately be used to describe the different types of sand. In such cases, the normal restrictions on noncountable nouns do not apply.

4. As noted above, special contexts sometimes alter this interpretation.

5. It should be borne in mind, however, that these categories depend heavily on culture-determined perceptions. For example, if we consider a concrete object like *fruit* in three languages like English, French, and Russian (each of which distinguishes between count and noncount nouns), we find that the latter two treat it as a count noun (Fr. *fruit*, Russ. *frukt*) while the former treats it as a noncount noun (*fruit*).

6. Of course, one can conjure up cases that seem to violate this generalization, e.g., "A most important man came to see me today." However, in such cases, the adjective's meaning is not interpreted as superlative, despite its form.

7. Generic interpretations are also found with the indefinite article ("A beaver has a flat tail and builds dams.") and with the article-less plural ("Beavers have flat tails and build dams."). Although the three types of generics have the same general meaning, they arrive at this meaning in different ways, each of which is consistent with the flow diagram. The definite article generic has already been accounted for, as focusing on the entire set of beavers. By contrast, the indefinite article generic randomly selects any individual member of the set, making that individual a representative of the entire set. The article-less plural is similar to the indefinite article case in that it randomly selects a subset to represent the entire set; in this case, however, the subset simply contains more than one member.

8. In case students raise questions about certain article choices in Figure 2 that are *not* tested—specifically, *the aid*, *the purchase*, and *interest*—you can explain that they are all part of "fixed phrases," or "idioms." The latter, for example, participates with *rate* as a compound form (note that it could even be hyphenated: *rate-of-interest*). Many such automatic article choices are found in English: *for the sake of, at the right time, in a hurry, for a while, with regard to*. (Note the impossibility of the following: *for a sake of, at a right time, in the hurry*. Since fixed phrases are not governed by rules, they will have to be learned individually.

Supplementary Reading

Many of the articles in this collection provide references on specific topics; however, we asked contributors to suggest additional appropriate readings. The following list is the result. Those interested in a more comprehensive bibliography might see Carolyn M. Blackman, "A Bibliography of Resources for Beginning Teachers of Technical Writing," *Technical and Professional Communication: Teaching in the Two-Year College, Four-Year College, Professional School,* ed. Thomas M. Sawyer (Ann Arbor, Mich.: Professional Communication Press, 1977).

Abelson, Herbert. *Persuasion: How Opinions and Attitudes Are Changed.* New York: Springer, 1959.

Baker, William H. "Teaching Business Writing by the Spiral Method." Paper presented at the First International Conference of the American Business Communication Association, 29 December 1975, Toronto, Ontario. ERIC ED 124 989.

Balsley, Howard L., and Vernon T. Clover. *Business Research Methods.* Columbus, Ohio: Grid, 1979.

Blicq, Ronald S. *Technically-Write! Communication for the Technical Man.* Englewood Cliffs, N.J.: Prentice-Hall, 1972.

Boettinger, Henry M. *Moving Mountains: The Art of Letting Others See Things Your Way.* New York: Macmillan, 1969.

Bowen, Mary E., and Joseph A. Mazzeo, eds. *Writing about Science.* New York: Oxford University Press, 1979.

Bruffee, Kenneth A. "Collaborative Learning: Some Practical Models." *College English* 34 (February 1973): 634–43.

Brunner, Ingrid, J. C. Mathes, and Dwight W. Stevenson. *The Technician as Writer: Preparing Technical Reports.* Indianapolis: Bobbs-Merrill, 1980.

Brusaw, Charles T., Gerald J. Alred, and Walter E. Oliu. *The Business Writer's Handbook.* New York: St. Martin's Press, 1976.

Burling, Robbins. "An Anthropological Glimpse of the English Teacher's World." *College English* 39 (September 1977): 18–24.

Chen, Ching-Chin. *Scientific and Technical Information Sources.* Cambridge, Mass.: MIT Press, 1977.

Corey, James R., and John E. Blodgett. *Content and Organization: Readings for Factual Prose Writers.* Boston: Holbrook Press, 1968.

Dodds, Robert H. *Writing for Technical and Business Magazines.* New York: John Wiley & Sons, 1969.

Ellman, Neil. "Peer Evaluation and Peer Grading." *English Journal* 64 (March 1975): 79–80.

Estrin, Herman. "An Engineering Report Writing Course That Works." *Improving College and University Teaching* 26 (1968): 28–31.

Fear, David E. *Technical Writing.* New York: Random House, 1973.

Fishbein, Morris. *Medical Writing: The Technic and the Art.* 4th ed. Springfield, Ill.: Charles C. Thomas, 1972.

Flower, Linda. "Problem-Solving Strategies and the Writing Process." *College English* 39 (December 1977): 449–61.

Flower, Linda. "Writer-Based Prose: A Cognitive Basis for Problems in Writing." *College English* 41 (September 1979): 31–49.

Gorman, Alfred H. *Teachers and Learners: The Interactive Process of Education.* 2d ed. Boston: Allyn & Bacon, 1974.

Gresham, Stephen. "Benjamin Franklin's Contributions to the Development of Technical Communication." *Journal of Technical Writing and Communication* 7 (1977): 5–14.

Halloran, S. Michael. "Teaching Writing and the Rhetoric of Science." *Journal of Technical Writing and Communication* 8 (1978): 77–88.

Harris, John Sterling. "So You're Going to Teach Technical Writing: A Primer for Beginners." *The Technical Writing Teacher* 1 (Fall 1974): 1–6.

Hawkins, Thom. *Group Inquiry Techniques for Teaching Writing.* Theory and Research into Practice. Urbana, Ill.: National Council of Teachers of English and ERIC Clearinghouse on Reading and Communication Skills, 1976.

Hays, Robert. *Principles of Technical Writing.* Reading, Mass.: Addison-Wesley, 1965.

Holtzman, Paul D. *The Psychology of the Speakers' Audience.* Glenview, Ill.: Scott, Foresman, 1970.

Hoover, Regina. "Experiments in Peer Teaching." *College Composition and Communication* 23 (December 1972): 421–25.

Houp, Kenneth W. and Thomas E. Pearsall. *Reporting Technical Information.* Encino, Calif.: Glencoe Press, 1977.

Krick, Edward V. *An Introduction to Engineering and Engineering Design.* New York: John Wiley & Sons, 1969.

Lannon, John M. *Technical Writing.* Boston: Little, Brown, 1979.

Lunsford, Andrea A. "Classical Rhetoric and Technical Writing." *College Composition and Communication* 27 (1976): 289–91.

Maltha, D. J. *Technical Literature Search and the Written Report.* New York: Elsevier-North Holland, 1976.

Mathes, J. C., and Dwight W. Stevenson. *Designing Technical Reports: Writing for Audiences in Organizations.* Indianapolis: Bobbs-Merrill, 1976.

McCarron, William E. "Confessions of a Working Technical Editor." *The Technical Writing Teacher* 6 (1978): 5–8.

Miller, Walter J. "What Can the Technical Writer of the Past Teach the Technical Writer of Today?" *IRE Transactions on Engineering Writing and Speech.* EWS-4 (1961): 69–76.

Mitchell, John H. *Writing for Technical and Professional Journals.* New York: John Wiley & Sons, 1968.

Moffett, James. *Teaching the Universe of Discourse.* Boston: Houghton Mifflin, 1968.

Morris, Charles J. "Simulation Evaluation Designs." Paper presented at the Annual Meeting of the American Educational Research Association, 4 April 1977, New York. ERIC ED 150 200.

Passman, Sidney. *Scientific and Technological Communication.* Elmsford, N.Y.: Pergamon Press, 1969.

Pearsall, Thomas E. *Audience Analysis for Technical Writing.* Encino, Cal.: Glencoe Press, 1969.

Pickett, Nell Ann, and Ann A. Laster. *Technical English.* 2d ed. San Francisco: Canfield Press, 1975.

Rathbone, Robert R. *Communicating Technical Information: A Guide to Current Uses and Abuses in Scientific and Engineering Writing.* Reading, Mass.: Addison-Wesley, 1972.

Rivers, William L. *Free-Lancer and Staff Writer: Writing Magazine Articles.* Belmont, Cal.: Wadsworth, 1972.

Sawyer, Thomas M. "Rhetoric in an Age of Science and Technology." *College Composition and Communication* 28 (December 1977): 390-98.

Sawyer, Thomas M., ed. *Technical and Professional Communication: Teaching in the Two-Year College, Four-Year College, Professional School.* Ann Arbor, Mich.: Professional Communication Press, 1977.

Schutte, William M., and E. R. Steinberg. *Communication in Business and Industry.* New York: Holt, Rinehart & Winston, 1960.

Sherman, Theodore A., and Simon Johnson. *Modern Technical Writing.* Englewood Cliffs, N.J.: Prentice-Hall, 1975.

Simon, Herbert. *Administrative Behavior: A Study of Decision-Making Processes in Administrative Organization.* New York: Free Press, 1957.

Snipes, Wilson. "An Inquiry: Peer Group Teaching in Freshman Writing." *College Composition and Communication* 22 (May 1971): 169-76.

Souther, James W., and Myron L. White. *Technical Report Writing.* 2d ed. New York: John Wiley & Sons, 1977.

Sparrow, W. Keats, and Donald H. Cunningham, eds. *The Practical Craft: Readings for Business and Technical Writers.* Boston: Houghton Mifflin, 1978.

Sullivan, Jeremiah J. "The Importance of a Philosophical 'Mix' in Teaching Business Communications." *Journal of Business Communications.* 15, no. 4 (1978): 29-37.

Ulman, Joseph N., Jr., and Jay R. Gould. *Technical Reporting.* New York: Holt, Rinehart & Winston, 1972.

Van Nostrand, A. D., C. H. Knoblauch, Peter J. McGuire, and Joan Pettigrew. *Functional Writing.* Boston: Houghton Mifflin, 1978.

Wells, Walter. *Communications in Business.* Belmont, Calif.: Wadsworth, 1977.

Contributors

Paul V. Anderson is Director of the Business and Technical Writing Program at Miami University. He serves on the editorial board of the *Journal of Technical Writing and Communication* and has edited an anthology for the Association of Teachers of Technical Writing.

Colleen Aycock is a doctoral candidate in Rhetoric, Linguistics, and Literature at the University of Southern California. She has taught technical writing courses in business, engineering, and health sciences at Northern Arizona University.

Ben F. Barton is Professor of Electrical and Computer Engineering at the University of Michigan. He has served as consultant for numerous government and industrial organizations. His publications include articles on teaching professional communications.

Marthalee S. Barton is Lecturer in Humanities, College of Engineering, The University of Michigan. Her publications include papers on technical communication and, as coeditor, "Technology and Pessimism," a special issue of *Alternative Futures.*

Charles E. Beck is head of the English Department at the United States Air Force Academy Preparatory School. Major Beck previously taught English, classics, and technical writing at the Air Force Academy and has served as a consultant in executive writing.

Renee Rebeta Betz earned the first master's degree in rhetoric granted by the University of Michigan and is a graduate student at the University of Illinois at Chicago Circle, where she hopes to receive the first doctoral degree awarded in that university's new rhetoric program. She has taught for the Cleveland Public Schools, Cleveland State University, and Central Missouri State University.

Anita Brostoff is Adjunct Assistant Professor, Education Center, Carnegie-Mellon University, and director of a secondary school writing project. She has served as codirector of the Communication Skills Center at Carnegie-Mellon and as consultant for business and secondary schools.

David L. Carson is Director, Masters Programs in Technical Writing and Communication, Rensselaer Polytechnic Institute, and serves as president, The Council for Programs in Technical and Scientific Communication, and vice-president, The Association of Teachers of Technical Writing. He is founder and director of the Technical Writing Institute for Teachers as well as cofounder of Sigma Tau Chi, honorary academic association of the Society for Technical Communication.

Gordon E. Coggshall is Manager of Marketing and Internal Communications at the Optical Technology Division of the Perkin-Elmer Corporation. He was formerly Assistant Professor and Coordinator, Applied Writing Program, Ithaca College. Among his publications are short stories and articles on writing and the teaching of writing.

Donald H. Cunningham is Professor of English at Morehead State University. He is a senior member of the Society for Technical Communication and has been editor for the Association of Teachers of Technical Writing since its founding in 1973. Among his publications are four books and a number of articles on technical communication.

Susan B. Dunkle is Director, Carnegie Institute of Technology Writing Center, Carnegie-Mellon University. She has served as communication consultant for a variety of businesses and industries and spoken and published on the consulting approach as method in the teaching of technical writing to engineering students.

Herman A. Estrin is Professor of English at New Jersey Institute of Technology. He has written many educational and scholarly articles and published eleven books on aspects of education, technical writing, and collegiate student affairs. He has received several awards for excellence in teaching, including the Distinguished Teaching Award by the New Jersey Council of Teachers of English and the Distinguished Newspaper Adviser Award by the National Council of College Publications Advisers.

Linda S. Flower is Associate Professor of English, Carnegie-Mellon University, and was previously director of the Business and Professional Writing Program, Graduate School of Industrial Administration. Her work includes articles on writing research and a recent textbook, *Problem-Solving Strategies for Writing.*

Stephen Gresham is Assistant Professor of English at Auburn University. He has served as a communications consultant and has authored numerous articles on technical writing and the history of technical communication.

Gerard J. Gross is a physicist for Locus, Inc., a systems engineering and electronics firm in State College, Pennsylvania. He previously taught technical writing and Shakespeare at the Pennsylvania State University. His publications include numerous technical reports as well as articles on literary analysis and writing.

Dean G. Hall is Assistant Professor of English at Wayne State University. He previously taught English at Kent State University.

Maurita Peterson Holland is Head, Technology Libraries, and Lecturer, Department of Humanities, College of Engineering, The University of Michigan. She is particularly interested in the integration of information resources into engineering education and practice.

Thomas N. Huckin is Assistant Professor of Linguistics, Department of Humanities, and Assistant Research Scientist, English Language Institute at the University of Michigan. He is cofounder and codirector of the University's annual summer conference on the teaching of scientific and technical English to nonnative speakers and coauthor of the forthcoming *English for Science and Technology*.

Lawrence J. Johnson is Assistant Professor of English at the University of Texas at El Paso. His publications include articles on writing programs for industry.

Peter R. Klaver is Associate Professor of English, Humanities Department, College of Engineering, The University of Michigan. His publications include articles on the use of simulation games in technical writing courses.

Wayne A. Losano is Associate Professor of English, University of Florida, and serves as Director of Freshman English and Coordinator of Business and Technical Writing. He has taught technical writing at Rensselaer Polytechnic Institute, Lowell Technological Institute, and Queensland (Australia) Institute of Technology.

William E. McCarron is Director, Freshman English, at the United States Air Force Academy. Major McCarron has published articles in *College English* and *College Composition and Communication* and is coauthor of a forthcoming book, *Persuasive Technical Writing.*

John H. Mitchell is Professor of English, University of Massachusetts, and president of the Association of Teachers of Technical Writing. An international consultant on the communication of scientific and technical information, he was a founding member of STC and is a fellow of ISTC (London), ATAIA (Sydney), and TCAA (Adelaide). As visiting professor, he has also taught at Arizona State University, Canberra College of Advanced Education, University of Hawaii, and the University of Victoria.

Leslie Ann Olsen is Associate Professor of Humanities, College of Engineering, The University of Michigan. She teaches rhetoric and technical writing, has developed computer-assisted instructional programs in writing, and is coauthor of two forthcoming texts on technical writing.

David M. Pahnos is Associate Director of the Carnegie Institute of Technology Writing Center at Carnegie-Mellon University. He teaches technical writing, serves as consultant for a variety of businesses and industries, and has published on the consulting approach as method in the teaching of technical writing to engineering students.

Thomas M. Sawyer is Professor of Humanities, College of Engineering, The University of Michigan. He has been a Fulbright Lecturer in English in Pakistan and a Visiting Professorial Lecturer at the University of Wales Institute of Science and Technology. Mr. Sawyer has published a number of articles, among them several on writing and speech.

Gretchen H. Schoff holds a joint appointment as Associate Chair of Integrated Liberal Studies and Professor of Technical Communications, University of Wisconsin—Madison. She also teaches a humanities course for the Institute of Environmental Studies. She has published papers on technical writing and interdisciplinary studies and has served as writing consultant to industries in the United States and Canada.

Dwight W. Stevenson is Professor of Humanities, College of Engineering, The University of Michigan. Specializing in technical and professional writing, he is founder and cochair of the annual Michigan Conference on Teaching Technical and Professional Communication. He also serves as consultant on technical and legal writing to numerous private industries and federal and state agencies. Among his books are *Designing Technical Reports: Writing for Audiences in Organizations, The Technician as Writer,* and *Problems in Exposition.*

William J. Wallisch, Jr., teaches composition, technical writing, and television at the United States Air Force Academy. As Director of Media Instruction, he helps cadets produce twice-weekly television programs over the Academy's closed-circuit system. Colonel Wallisch's articles on television production and education have been published in various academic and trade journals; last year his paper on teaching through television won the outstanding paper award at the International Technical Communication Convention.